CAITLYN'S CHRISTMAS

LAURA SCOTT

Caitlyn Weston gasped and slammed on the brakes as a white animal streaked across the road. Wrenching her small car over onto the shoulder, she immediately pushed open her driver's side door to climb out, despite the blustery December wind.

The animal had been a small cat, and she was fairly certain she'd glimpsed blood marring the side of his or her coat. As a veterinary technician, she adored animals and couldn't bear the idea of this small cat being out in the cold, especially while injured. She glanced over her shoulder nervously as she quickly headed into the woods where the cat had disappeared. Her hometown of Sevierville, Tennessee, which was nestled near the Smoky Mountains, was only five miles away. Still, it was late, just after midnight, and the moon and the stars were hidden behind thick clouds.

"Here, kitty. Nice, kitty." The white cat might be feral, although she found it strange that it would be running around this far out of town where it would be more likely to end up as a coyote's meal ticket.

"Kitty? Nice kitty?" she called again. Patches of snow lingered from a storm a few days ago, making it difficult to see the white cat.

There! Light-colored eyes winked at her from the darkness.

"Here, kitty." She wished she had food to entice the animal closer. While working at the veterinary clinic, she always had treats in her pockets. Tonight, though, she'd been out on a date, a failed experiment since Nate Powers, the guy she'd been with, had been far more interested in the female lead in the country-western band they'd gone to hear. She doubted he'd even noticed she'd left.

"I'm not going to hurt you," she murmured, edging closer. "I just want to help. You're cold and hungry, right? I can help you."

The cat blinked but didn't move. Caitlyn came a little closer, frowning when she realized the feline was shivering. She unwrapped the red scarf from around her neck and held it out. "Here, kitty. Do you want to warm up?"

She'd been told by Dr. John Vice, one of the veterinary docs at the clinic, that she had a voice animals responded well to. She continued inching toward the cat when the ball of fur stood and came over to nose the scarf.

Yes! Caitlyn slowly and carefully folded the other end of the red scarf around the cat. The feline burrowed into the scarf, seeking warmth and comfort.

Definitely not a stray, she thought as she gathered the cat and the scarf up from the ground. Cradling the animal to her chest, she wondered if the cat was microchipped as she turned to head back to her car.

A muffled scream stopped her dead in her tracks.

What was that?

Caitlyn turned, raking her gaze over the wooded area.

Some wild animals could sound like people, especially while mating. Was that what she'd heard? She took another step, then she heard a deep male voice.

"You're going to pay, not me!"

"No, please—"

A harsh slapping sound made her gasp. Caitlyn shivered as repressed memories of the time she'd lived with the Preacher flashed in her mind. The Preacher had slapped her across the face when one of the older foster kids had stood up to him in a way he deemed unacceptable. He'd liked using her as a pawn to control her older foster siblings.

But that was a long time ago. She was safe now.

Yet another woman wasn't.

She needed to do something. She had her phone, but she was worried that the man might hear her make a 911 call. Maybe she could take him by surprise, give the woman enough time to get away. Yes, that was a good plan.

But first she had to find them.

Easing through the woods, she made her way toward the muffled sounds of someone struggling.

Movement between two thick trees caught her attention. The cat in her arms began to purr, making her wince. Was the sound loud enough that the arguing couple could hear? She made her way through the brush as silently as possible. Where were they? Finally, she was close enough to see them. A tall man wearing a black coat was using his large hands to choke a woman with long dark hair. Her heart thundered in her chest as she watched in horror. How long had he held her like that? How long had it taken her to get through the woods? Too long. The woman's entire body was slack as he continued cutting off the circulation to her brain.

No! Caitlyn fumbled for her phone but then froze when

the woman dropped in a crumpled heap to the ground. Even from here, she could see the whites of the woman's open eyes, staring up at the dark sky. The man glared down at her for long moments, breathing heavily, before mumbling something under his breath and turning away.

Run!

Caitlyn sensed there was nothing she could do for the woman now. She'd been too late. The woman was dead.

And the man could easily kill her too.

Shrinking backward, she retraced her steps, trying to remember where she'd left her car. The sense of urgency propelled her forward, and she feared she was making too much noise. That the man who'd just murdered that poor woman would realize she'd seen him.

The cat in her arms was silent now, maybe realizing the precariousness of her position. Tossing a furtive glance over her shoulder, she nearly tripped over a log half hidden in a patch of snow.

When she burst through the trees, Caitlyn nearly sobbed in relief when she saw her small gray Honda. She ran toward it, yanking the door open and sliding in behind the wheel. Keeping the cat on her lap, she started the car and gunned the engine. Peeling away from the side of the road, she didn't have time to relax. The moment she drove around the bend, she saw the dark pickup truck. And the man striding purposefully toward it.

Her headlights flickered over him, and she caught a better look at his face.

And recognized him. Not his name, but his face. The guy had been at the tavern called Flannery's where she'd met up with Nate to hear the country band. For a split second, their eyes locked before she passed him.

She waited until the man and the truck had disappeared behind her before she pulled out her phone with trembling fingers. Caitlyn quickly dialed 911 but then disconnected before the operator had a chance to answer.

What was her emergency? The woman was dead. The man had killed her. Mesmerized by the man's familiar features, she hadn't even looked at the license plate of the truck.

She inwardly railed at herself for being so stupid. She'd done nothing to help that poor woman, couldn't even find a way to track the man who'd hurt her.

Then again, she hadn't expected the guy to kill the woman. Sure, he'd slapped her, but to kill her? Who did that?

She shivered again despite the heat blowing from the vents. A murder. She'd witnessed a terrible, brutal murder.

The whole thing seemed surreal. Like maybe she'd imagined the entire event. Only she knew she hadn't.

The cat purred again, and she glanced down at the feline's blue eyes. If not for going after this injured kitten, she wouldn't have been there. She struggled to calm her racing heart and thumbed through her directory to find the only cop she knew by name. The man who'd kept her and her sister safe from harm two months ago.

Devon Rainer.

THE RINGING PHONE jarred him from sleep. Twelve forty-five in the morning? Who was calling him at this hour? Grabbing his phone, he blinked the sleep from his eyes to peer at the screen. He'd thought maybe his partner, Bruce Whitmore, was calling, the guy was always arguing with his wife about something, but it was a different name on the screen. He bolted upright in bed. "Hello? Caitlyn? Is something wrong?"

"I-I'm so sorry it's so late, b-but I didn't know who else to call." Her voice was hoarse as if she'd been crying.

"Are you hurt?" A myriad of possibilities of what may have happened flashed through his mind. "Tell me where you are, I'll come meet you."

"I'm not hurt." She sniffled loudly. "But I saw something terrible. I'm heading toward my apartment, will you please meet me there?"

"Yes, how far away are you?" He pinched the phone between his ear and his shoulder as he pulled on his jeans. Then he reached for his navy blue Sevierville PD sweatshirt. "I can be there in five minutes."

"Thank you. I'll be there in about that time too."

He was relieved she wasn't injured, but he was hesitant to disconnect from the call. "Do you want me to stay on the line with you?" Technically, she shouldn't be talking on the phone while driving, but this sounded like an exception to the rule. What terrible thing had she seen?

Meow.

He frowned, still holding on to his sweatshirt. "Is that a cat?"

"Yes, she's wounded. And no, you don't have to stay on the line. I'm only two miles from my apartment building, so could you please hurry?"

"Sure, be there in five." He disconnected from the call, shoved the phone into his pocket, then pulled on his sweatshirt. As Caitlyn was a veterinary technician, he hoped the terrible thing she'd seen wasn't just that someone had hurt the cat.

Like most people, he abhorred animal abuse. There was no reason to hurt a pet. Yet being woken after midnight for such an event seemed a bit extreme. But maybe not to Caitlyn. He knew how much she cared about animals.

Devon pulled on his coat, grabbed his keys, and climbed into his SUV, which was parked in the attached garage. Thanks to the late hour, he made it to Caitlyn's apartment in less than the promised five minutes.

She was waiting for him in her dark gray two-door Honda. When she saw him, she pushed open her door and stood, holding a balled-up red scarf to her chest.

No blood on her face and no evidence of a car crash, so that was good. He crossed over to her. Up close, he could see the small white face of a cat with piercing blue eyes, eerily similar to Caitlyn's, peering out from beneath the red scarf. "What happened? Did someone hurt the cat?"

"Huh? Oh, yes, maybe. I haven't had time to examine her injury." Caitlyn shivered. "Can we go inside to talk? My roommate, Annette, is out of town for two weeks visiting her parents in Florida for the holiday." She glanced around the desolate parking lot. "I'd rather not stay out here."

"Of course." Devon knew Caitlyn well enough to understand she wasn't inviting him in for anything personal. Two months ago, Lincoln Quade had asked him to help protect Caitlyn and her sister, Jayme. At the time, Caitlyn had seemed interested in him. He'd kept his distance, first because he needed to be professional, but more so because despite being drop-dead gorgeous, with her long blond hair and big blue eyes, she was too young.

Barely twenty-three to his twenty-eight.

Devon had been involved with Sabrina for three years before she'd left him for another man. One who wasn't a cop because she'd suddenly decided police officers were terrible people riding some big power trip when it came to dealing with the public. *Thanks, media,* he thought sourly.

Since then, he'd kept his relationships light and fun. Easy peasy. No promises of forever.

He'd sensed Caitlyn was a bit naïve when it came to relationships with men. During the short time they'd spent together, he'd learned a little about her past. How she'd spent a couple of years in an abusive foster home before escaping with her older foster sister Jayme thirteen years ago. He'd understood Jayme had protected Caitlyn the best she could from the harsh realities of living on the streets. Admiring what Jayme and Caitlyn had done was easy.

They were both smart and beautiful. But they also were big believers in God, which wasn't really his thing.

"Will you unlock the door for me?" Caitlyn's voice cut into his thoughts.

He took the keys from her shaky hand and did as she'd asked. He held the door for her, then followed her up to her apartment where he unlocked that door too. There were Christmas decorations scattered around the room, including a large fake tree in one corner of the room. It appeared as if Caitlyn was already well into the holiday spirit.

"Thank you." Caitlyn sank down onto the sofa, letting out a long sigh. She shifted the scarf-wrapped cat to the cushion next to her, then ran her fingers through her hair. Her face was pale, her gaze cloudy with fear. "Snowball has a laceration that needs attention, but this is more important."

Snowball? He was tempted to smile, but the devastated expression on her face held back the mirth. At least the horrible thing she'd seen didn't involve the cat. He chose to sit on the chair closest to her, reaching out to take her cold hand in his. "What happened, Caitlyn? Why do you look as if you've seen a ghost?"

She stared at their entwined fingers for a moment

before meeting his gaze. "It was worse than that. So much worse."

The tiny hairs on the back of his neck rose in alarm. "Tell me."

She drew in a deep breath. "I, uh, was driving along highway double G when Snowball streaked across the road in front of me. I caught a glimpse of blood staining her white coat, so I pulled off the road and went looking for her."

He nodded encouragingly. "Apparently, you found her."

"I did. But then I heard arguing between a man and a woman." Her fingers tightened around his. "She was pleading with him, then I heard a slap."

He winced. "Where were they?"

"In the woods." She shook her head. "I know this sounds crazy, but it gets worse."

Worse? In the six years he'd been on the job, he'd seen a lot of bad stuff. People could be so cruel to each other. He forced himself not to overreact. "Go on."

"I, uh, couldn't see them. So I moved farther into the woods. I thought maybe I could interrupt their argument, give the woman time to get away."

Dread coiled in his gut. "Did you find them?"

"Yes." Her voice dropped to a whisper. "But it took me a while to get close enough to see them, and when I looked through the trees, he was . . . strangling her."

"Strangling her?" He couldn't help but echo what she'd said. "Was she able to get away?"

"No. He must have been choking her for a while because her entire body was limp, but he didn't stop. Not for what seemed like forever. When he finally let her go, she fell to the ground in a heap. Even from a distance, I could see that her eyes were open and staring blankly up at the sky." She paused, gulped, then whispered, "He killed her."

"Did you call 911?"

Caitlyn shook her head. "I was going to once I got back into my car. I started driving. When I came around a curve, I saw the man walking toward his pickup truck. And I recognized him."

"You what?" He jumped to his feet. "Who is he? We need to issue a BOLO for him as soon as possible!"

"No, I don't know his name. I recognized him from Flannery's. He was there tonight, the same way I was." She lifted her tortured gaze to his. "And I think he might have recognized me too."

The chill hardened to ice. "He saw you?"

She gave a jerky nod. "I think so."

He raked his hand through his dark hair and began to pace. "There must be a way to find out who he is. Did you get his license plate number?"

"No." Tears welled in her eyes. "I'm sorry, I know I messed up. But it all happened so fast . . ."

"Shh, it's okay." He reached down to pull her up into his arms. "You didn't mess up. I'm just glad you were able to get away without being hurt." He didn't even want to think about how close Caitlyn had come to being strangled the way the other woman had been.

She buried her face against his chest. After a few minutes, though, she pulled herself together. "I need to look at Snowball's injury and give her food and water."

He glanced down to where the cat was staring up at them. "Okay, but we need to go back to that spot in the woods. Do you think you can find it?"

She stiffened. "Go back? Tonight?"

"Yes." He hated the stark fear in her gaze. "I need to get

the local cops out there too. And a detective. But I think we should find this poor woman first. Keep her safe from predators."

Caitlyn paled but reluctantly nodded. "I just need a few minutes."

He wanted to tell her to leave the cat until they returned, but knowing Caitlyn, she'd refuse. And since he didn't know how long they'd be gone, he let it go.

The dead woman couldn't get any more dead.

True to her word, Caitlyn didn't take long. She carried the cat into the kitchen, talking softly to the animal as she cleaned out her wound. Then she opened a can of tuna and emptied it into a shallow dish. After filling another shallow dish with water, she stood back, watching with satisfaction as the cat began to eat and drink.

"I still need a litter box," she said. After a quick glance around the room, she went over and gathered several potted plants, including a few orchids. She yanked the plants out and tossed them into the garbage. She emptied the containers into a shallow rectangle-shaped pan. She dumped the dirt in first, then layered the bark chips along the top. "It's not kitty litter, but it will have to do."

He lifted a brow. "Won't Annette be upset about the dead plants?"

Caitlyn shrugged. "I'll buy new ones, besides, if Annette was here, she'd care more about Snowball than a couple of stupid plants."

He suspected she was right. Her roommate was also a veterinary tech at the local veterinary clinic. She was a sweet, attractive kid too. But, for some reason, he'd been drawn to Caitlyn.

Too young, he reminded himself firmly. *And far too innocent.*

Caitlyn turned and crossed over to the apartment door. "Let's go."

He followed her out, waiting as she locked the door before they went outside. Cupping her elbow in his hand, he steered her toward his SUV. Glancing at her small Honda, he thought it was a minor miracle that she was able to stay on the twisty mountain highways in that tin can of a car.

"Head toward Knoxville on highway double G," she said once they were settled inside. "This happened about five miles outside of Sevierville."

Five miles might be outside their jurisdiction, but he didn't know that for sure. He hoped the area didn't belong to the park rangers. He cranked the heat and pulled out of the parking lot of her apartment building. A few turns later, they were headed out on highway double G.

He went slow, more so to give Caitlyn the opportunity to recognize the location where this had all taken place. With trees and brush lining both sides of the road, he was concerned that one section of the woods would look just like another.

Would she be able to pinpoint the spot? He hoped so, otherwise he'd be forced to bring half the police department out to search for the poor woman, or what might be left of her, in the morning.

"Slow down," Caitlyn cautioned as she peered past him. He thought they may have to go a few miles out of their way and turn around so that she would be able to see more clearly. "Stop!"

He hit the brakes, harder than he intended, bringing the SUV to a rocking halt. "Where?"

"There, see that small patch of snow in front of that

birch tree?" Caitlyn gestured impatiently. "That's where I went into the woods to find Snowball."

He tried not to let his doubt show on his features. There were many birch trees and several patches of snow. But he had little choice but to give her the benefit of the doubt. "Okay, let's check it out."

She readily jumped out of the car, heading toward the spot she'd indicated. He let her take the lead, hoping they wouldn't get lost.

The path she took was a zigzag pattern. Several times he wanted to ask if she knew where she was going but managed to bite his tongue. She stopped and pointed to a spot near an evergreen tree.

"That's where I found Snowball." She glanced at him. "And where I first heard them arguing."

He swept his gaze over the area. "She's somewhere close by?"

"Not that close. This way." Caitlyn turned and continued walking.

He noticed a few footprints in the snow, indicating she had indeed found the correct location. He wanted to ask how she knew, but she quickened her pace.

"Caitlyn?" He hurried to catch up. She stood in a small clearing, looking frantically around. "Here. She should be right here. Where is she?"

He reached out to grasp her arm. "Calm down, maybe she wasn't dead. Maybe she fell unconscious and managed to get out of the woods on her own."

"No, I saw her eyes. He cut off her air for a long time. I'm telling you, she was dead." Caitlyn waved her arm impatiently. "This is the place it happened, Devon. I know it is."

"I believe you." He blew out a breath. Caitlyn had found this location without any hesitation.

She also claimed she'd witnessed a murder. Yet without a suspect, or a body, there was no proof that a crime had been committed here at all.

Devon scowled. He'd still notify his boss and get a detective assigned to the case, but he had a bad feeling about this.

He was very much afraid this guy might come after Caitlyn next.

CHAPTER TWO

It was too dark for Caitlyn to read Devon's facial expression. Did he really believe she'd seen a brutal murder? Honestly, she could understand his skepticism. She found it difficult to believe too. But she knew she hadn't imagined what had happened. She never drank alcohol, another fact her so-called date hadn't appreciated, so her mind had been clear. She'd heard the fighting, made her way through the woods in time to watch the guy strangle the woman.

Only now the body was gone.

She focused on the area where the crime had taken place. The ground wasn't disturbed, the way it would be if her body had been dragged away. Leaving her to believe the guy must have gone back into the woods and carried her out.

Because when she'd come around the bend, he'd recognized her. And he'd known she'd witnessed his crime.

A chill snaked down her spine.

Would he be able to figure out who she was? She swallowed hard against a surge of panic. What if he went back to Flannery's and asked around? It wasn't as if she was a regular there, but Nate was. She didn't doubt Nate would blab her full name to anyone who bothered to ask.

"Caitlyn?" Devon came over to lightly touch her arm. "Are you okay?"

"No." She struggled to draw a deep breath. "I'm sure after he saw me and realized I'd witnessed the crime, he went back to carry the woman out of the woods."

"This isn't your fault, Caitlyn. It's his. We need to stay back until the crime scene techs and the detective assigned to the case get here." He subtly tugged her back. "Maybe they'll find his footprints in the snow."

"Maybe." She allowed Devon to pull her back the way they'd come. It had been many years since she and Jayme had lived in the woods after escaping the nightmare of the Preacher, but she still knew the basics of tracking. The cold December ground would be too hard to yield prints, but the snow may have captured some impressions. Unless the murderer had purposefully avoided those areas.

Exactly what she would have done in his shoes.

She listened as Devon spoke with the police dispatcher. His calm, deep voice helped soothe her fears. There was nothing to worry about. This guy wouldn't come looking for her. Without a body, he had to realize there was nothing to back up her wild story.

Right? Right.

"Detective Wade Ernest is on his way," Devon said, breaking into her troubled thoughts. "He'll want your statement."

"I know." She understood how law enforcement operated. She only hoped this Detective Ernest would believe her. The way Devon seemed to.

"Hey, it's okay." Devon looped his arm loosely around her shoulders. "I'll keep you safe."

"I know you will." Devon had protected her two months ago, leaving her impressed with his abilities. He was an attractive guy, and at one point, she'd hoped for something more, but he'd held her at arm's length, refusing her rather lame offer to meet up for coffee sometime.

She told herself that it was for the best. After her recent experiences with dating, both attempts had ended badly, she wasn't really interested in trying again. Frankly, she'd always known her strength was relating to animals more so than people.

Like Snowball. On Monday, she'd have to take the cat into the clinic to check if she was carrying a microchip. If not, then she'd need to put up notices on social media in an effort to find the animal's owner. No matter how much she'd love to keep the pretty kitten, she couldn't do that if it belonged to someone else.

Bright headlights pierced the darkness. She reluctantly pulled away from Devon, turning to face the officers who'd responded to his call.

"Rainer? What's this about a murder without a body?"

She narrowed her gaze at the detective who approached them. He looked irritated at being woken up, and she wondered if they should have just waited until morning. After all, there wasn't a body to protect from predators, was there? Still, this was a serious crime, so she tipped her chin and met his gaze. "Detective, my name is Caitlyn Weston, and I'm the one who witnessed the murder."

"Weston?" Detective Ernest eyed her thoughtfully. "You're Jayme's younger sister, right?"

"Yes." Her foster sister had recently married the local arson investigator, Lincoln Quade. Linc knew a lot of the cops by name, and obviously several of them knew Jayme now too. It was nice to have the police on their side rather than hiding from them the way they had when they'd first escaped the Preacher. "They were over in that clearing when it happened. He strangled her, then let her drop to the ground. He left, but when I saw him on the road next to his pickup truck, he must have realized I witnessed the crime and decided to hike back into the woods to remove the body."

The detective glanced from her to Devon and back again. "Okay, I think you need to start at the beginning. Come with me, please." Detective Ernest escorted her to his vehicle so he could take notes. While they talked, Devon walked the perimeter of the crime scene with the other officers and two crime scene techs who'd responded to his call.

To his credit, Ernest didn't interrupt. He waited until she went through the entire story before clarifying a few things. "You were at Flannery's with a man named Nate Powers, but you left on your own?"

"Yes." She shrugged. "He was pretty focused on the lead singer of the band that was playing. I doubt he noticed me leaving."

"And you're absolutely sure you saw the man and the woman at Flannery's before the murder."

"Not the woman," she corrected. "I didn't recognize her at all. But the guy was there. I only remember because I was kinda bored and spent a fair amount of time watching the people at the bar." She brought up the brief image in her head. "The guy was in the back right corner, standing against the wall. I noticed he wasn't paying attention to the band either, and our eyes met for a brief instant. I looked away because I didn't want him to think I was interested or anything, because I wasn't." She twisted her fingers together in her lap. "But when I saw him standing near the pickup truck at the side of the road, I know he recognized me too."

"I'll have officers canvass Flannery's; I'm sure someone there will know his name." Detective Ernest frowned. "Although all we can do is question him. Without a body, we can't verify your story."

"I know what I saw," she insisted. "Why would I make up something that awful?"

"I'm not saying you would, but it's your word against his."

"You need to find out if there are any recent missing person reports for a woman with long dark hair." She stared at Ernest over the center console. "She isn't a figment of my imagination. When you find this guy's name, maybe you'll also discover his wife, fiancée, or girlfriend is missing." She found herself sending up a silent prayer that the woman hadn't been a complete stranger to the man who'd killed her, or they might never figure out who she was.

The detective nodded. "I hope it's that easy. In the meantime, you need to take extra safety precautions. Don't go anywhere alone, make sure you lock your doors, that kind of thing."

She couldn't help glancing over at Devon. "My roommate is out of town for the next two weeks. I live alone and have to go to work on Monday morning."

Ernest grimaced. "Maybe you should stay with your sister for a few days. Just until we have a chance to find and question this guy."

"I can't because Jayme and Linc are on their honeymoon." It was a trip they'd been forced to cancel prior to Thanksgiving because of a large apartment fire in Knoxville. Linc had been called in to investigate, eventually finding and arresting the person responsible. Since it was early December by then, Jayme and Linc had waited until her graduation from her associate degree program before finalizing their trip. This past Friday had been Caitlyn's graduation ceremony, making her a full-fledged veterinary technician. Early Saturday morning, Jayme and Linc had packed up to take their belated honeymoon. Annette had left around noon too; her graduation gift was a trip home to see her parents in Florida.

Being alone without Annette or her sister to hang out with had been one of the reasons she'd agreed to meet Nate at Flannery's. Something she wished she'd declined.

Especially since Jayme and Linc wouldn't be back until next Sunday.

"Detective?" Devon rapped on the window, startling her. Ernest rolled down the window. "The techs have found a partial print in the snow. You might want to take a look for yourself."

"I do." Ernest pushed open his driver's side door to slide out from behind the wheel. Caitlyn got out of the car too.

"Can I see it?" she asked.

"No, we should stay here." Devon glanced down at her hiking boots. "You were wearing those earlier, right?" She nodded. "They'll want your boot print for a comparison."

"Okay." She stared down at her feet. Hmm. Maybe the fact that she hadn't bothered to dress up for her date had contributed to Nate's ignoring her. She preferred casual clothes like jeans, sweaters, and hiking boots. It hadn't occurred to her to dress to impress.

Whatever. She didn't care about Nate. In hindsight, it was a good thing she'd left Flannery's when she did.

The next thirty minutes passed with excruciating slowness. She was desperate to get home, get warm, and cuddle with Snowball for the short time she'd be able to. Thankfully, the laceration along the cat's side wasn't too deep.

Since she didn't know when the cut had happened, suturing it closed wasn't an option.

But she may need to use the cone of shame or one of their newfangled soft collars to prevent Snowball from licking the wound.

Worrying about the cat was easier than worrying about her own safety. Should she ask Devon to stick around for what was left of the night? Or would he take her request as something more personal?

She really wished she hadn't tried to ask him out. In her usual clumsy way.

"Caitlyn? Come with me." Devon waved her over.

"Where?" She frowned, joining him at the side of the road.

"The boot print is around the curve." There was an excited glint in his eyes. "Near the spot where we believe you saw him standing near his truck."

They walked quickly along the edge of the road, stopping a few feet from where an officer and a tech stood. There were tiny orange cones marking the ground.

"This is where I saw him," she confirmed. "And I didn't get out of the car here, so that print can't be mine."

"It's not yours, it's a size twelve shoe." Devon drew her closer. "Can you see it?"

The tech held a flashlight pointed down at the print. Detective Ernest stood nearby, watching her. "I can see it," she agreed. "Looks like it's a hiking boot, similar to the brand I'm wearing." Hers were a size eight, not a twelve.

"Yeah, we noticed that." Devon looked at her. "Do you remember seeing his feet? Or what he was wearing?"

"Not his feet, but he was wearing a black jacket." She felt like a complete failure. "Sorry, but I was focused on the expression of surprise on his face when he saw me."

"It's okay," Devon hastened to reassure her. "I had to ask."

She backed away from the partial print. "How much longer do we need to stay?"

Devon glanced over at the detective who nodded. "I'll take you home."

"Thanks." She would be glad to get away from this area. Why had the guy chosen this place to murder the woman? Had they been fighting in the car? Maybe she'd gotten out and walked into the woods to cool off, but he'd followed.

She walked alongside Devon to his SUV. Neither of them said anything as he drove back to her apartment. When he pulled in next to her Honda, she unlocked her seatbelt. "Thanks for the ride."

"Wait." He shoved out of the car and ran around to open her door. "I'll walk you in."

She appreciated his gesture. They took the stairs to her second-floor apartment. When she used her key to unlock the door, he followed her inside.

"I was thinking I should sleep on the sofa tonight."

"Really?" She turned to face him, hoping her relief wasn't too evident. "I can also change the sheets on Annette's bed, she won't be here for two full weeks. I'm sure she wouldn't mind you staying."

A smile quirked the corner of his mouth. "I guess that means you aren't too upset by my suggestion."

"I know it's a huge inconvenience for you," she said softly. "But I'll sleep better knowing you're nearby."

"It's not a problem," he assured her.

Caitlyn wanted to hug him but forced herself to turn away. Devon was simply being a nice guy and a good cop.

His offer to stay was likely a way for him to stay in

Linc's good graces. And honestly? Right now, she didn't care why he'd agreed to stay.

Maybe she was being a big wimp. But she'd gladly take what she was given rather than staying in the apartment all alone.

DEVON SAT ON THE SOFA, waiting for Caitlyn to finish in Annette's room. He was glad she hadn't refused his offer to stay, or he'd have been stuck sleeping out in his car, which would have been really cold.

Snowball padded over to rub her face against him.

"Hi, kitty," he said, surprised that the feline let him pet her. "I hope you didn't make a mess for Caitlyn."

Meow.

He didn't understand cat speak, yet it seemed as if the animal had made herself at home. Trust Caitlyn to take off after a stray animal only to end up witnessing a murder.

It was concerning to think this guy might find out who she was and come looking for her. Devon trusted Detective Ernest's abilities, once the guy got over being woken up in the middle of the night. Finding the boot print had given credence to Caitlyn's story. And there was the fact that Caitlyn was Linc's sister-in-law. Ernest likely figured he needed to take Caitlyn's story seriously.

Snowball jumped up onto the sofa and crawled into his lap. She curled into a ball and purred again.

"Are you sure you didn't lose a cat?" Caitlyn asked wryly as she walked into the room.

"Not me, sorry."

She sighed. "I hope I can find her rightful owners." She paused, then added, "Thanks again for staying."

"I'm happy to do it." He gently eased the cat off his lap. "Try to get some sleep."

"I will. Good night, Devon." She gathered Snowball into her arms and went into her room, closing the door behind her.

He poked his head in Annette's room but knew he wouldn't be able to sleep. At least, not yet. He took off his shoes and padded from window to window, peering outside to make sure no one was lurking nearby and double-checking the locks. He'd wanted to head over to help canvass Flannery's, but Ernest wouldn't let him because he was off duty and apparently too close to the case. Or rather, too close to Caitlyn.

Glancing at his watch, he realized it was now close to three in the morning. If he didn't get some sleep soon, he'd be useless tomorrow.

Devon forced himself to stop pacing. He shut off the single kitchen light but left the Christmas tree lights on before heading into Annette's room. Leaving the door open about an inch, he stretched out on the bed fully clothed, except for his shoes. Not just because he didn't want Caitlyn to see him half dressed but because he might need to respond to a threat at a moment's notice.

Logically, it didn't make sense that the murderer would show up at Caitlyn's apartment building. Not this fast. He'd have to find out her name and address first, no easy feat.

No, the odds were against that happening. There was no reason for him to be tense. Devon did several deep breathing exercises to relax his muscles.

Now, if only he could relax his mind. Being this close to Caitlyn made it difficult to remember why he'd chosen to keep her at arm's length. Oh yeah, her innocence, at least when it came to men.

Her life experiences, though, were far from innocent. Especially now. He hated knowing she'd witnessed a brutal crime.

And felt honored that she'd come to him for help.

Devon must have dozed because he awoke to the enticing scent of coffee. He rolled off the bed and quickly washed up in the bathroom before padding out to join Caitlyn.

"Good morning, Devon." She was sitting at the tiny kitchen table, holding Snowball on her lap. If anything, Caitlyn looked even more beautiful, despite her tangled hair and lack of makeup. He had to force himself to look away before he made a complete fool of himself. Idly, he noticed there was dirt on the floor from where the cat had used the makeshift litter box.

He cleared his throat. "Good morning. Mind if I have some coffee?"

"Of course not, please, help yourself. I have peppermint-flavored creamer, too, if you'd like some."

He tossed her a wry grin. "Don't you know that all cops drink their coffee black?"

She laughed, a light musical sound that made him smile. "I'd never survive being a cop. I don't even like coffee as much as I love the flavored creamer. The coffee is just an excuse because, you know, it would look weird to just drink the creamer."

He chuckled as he filled his mug. "I hope you slept okay."

"I did, thanks." She eased the cat off her lap, gently setting Snowball on the floor. "I have eggs and cheese, do you like omelets?"

"I love them, but you don't have to cook for me." He

didn't want her to feel obligated to feed him. "I can grab something on my way home."

There was a flicker of fear in her clear blue eyes before she turned away. "It's no trouble, I was going to make one for me anyway."

He could tell she didn't want him to leave, at least, not yet. "Okay, thanks."

She busied herself at the stove, cracking eggs into a bowl, then beating them with a whisk. "I usually attend church services on Sunday mornings. Are you interested in joining me?"

He nearly choked on his coffee. "Uh, no thanks. I plan to check in with Detective Ernest, see if he has any other leads."

She glanced at him over her shoulder. "It sounds like you don't believe in God."

"It's not that," he protested, pushing away from the kitchen counter to sit at the table. "I just never learned much about religion."

"So, you do believe in God?"

Her persistence was wearing on him. "I don't know, maybe." He took another sip of coffee. He hadn't used the peppermint creamer, but there was a hint of flavor in his coffee anyway. Or maybe the scent lingered in the air. He had to admit, he kinda liked it.

"I didn't believe in God either, until recently," she said, turning her attention back to making the omelets. "Jayme and I escaped from a man who we thought of as the devil, and he claimed God was angry and was punishing us for being sinners. It wasn't until we met Linc that we learned God wasn't like that at all."

Linc? Believed in God? He couldn't deny being surprised. "Really?"

"Yes, he said God was always there, watching over us." She poured the egg mixture into the skillet and glanced at him. "I know how you feel because going to church and praying was uncomfortable at first, but now I'm glad to have God's shining light and His guidance."

He didn't know what she expected him to say, so he remained silent.

A few minutes later, the omelets were ready. She brought the plates over to the table, setting one in front of him.

"Looks great, Caitlyn. Thanks."

"You're welcome." She bowed her head and said, "Lord, thank You for keeping us safe in Your care and please bless this food we are about to eat. Amen."

He wasn't sure God had anything to do with keeping them safe, but he wasn't going to argue. The food was delicious, and he enjoyed every bite. When he was finished, he carried his plate over to the sink. "What time are you going to church?"

She looked at him in surprise. "Service starts at ten. I usually like to be there a few minutes early."

"I'll follow you to church, make sure you get inside safely." The offer popped out of his mouth before he could prevent it.

"Thanks, but I'm sure I'll be fine." She finished her omelet and joined him at the sink. "Would you do me a small favor, though?"

Her nearness was disconcerting. "What kind of favor?"

"Will you let me know when they find this guy?" Her blue gaze held his. "I'm going to be constantly looking over my shoulder until they do."

"Yes, I'll make sure you're kept informed." He took a

step back to avoid succumbing to the temptation to kiss her. "Detective Ernest will want you to ID the guy."

Her gaze clouded, but she nodded. "I'll identify him and testify against him as needed. What he did to that poor woman . . ." Her voice trailed off.

"Don't," he urged, putting a hand on her slim shoulder. "Try not to think about it."

"I'm surprised I was able to get any sleep," she whispered. "It was difficult to push the images out of my mind."

His resistance melted, and he drew her into his arms. "I'm sorry you had to see that, but I'm sure they'll find and arrest him soon."

"I hope so." Her voice was muffled against his T-shirt. Then she surprised him by pulling away. "Well, I need to shower and get dressed. You can use Annette's shower, too, if you like. Excuse me." She hurried out of the kitchen, leaving him staring after her.

He didn't have a change of clothes but decided to make use of the shower anyway. As he did, he found himself thinking about what Caitlyn would do after church. Hang out here in the apartment with a small white kitten for company?

With a groan, he realized he couldn't do it. He couldn't just let her go to church, then sit here all day on her own.

He'd go with her, even if that meant attending church. Something he hadn't done since he was a small boy, before his parents had gotten divorced. They'd fought over him, the house, the car, and just about everything else on the planet, which had always made him think that going to church was something they'd only done for show. And not because either of them had truly believed in God.

Whatever. He had the day off, why not spend it with

Caitlyn? All he needed to do was remind himself that she was a friend.

When Caitlyn emerged from her room wearing a pair of pale gray dress slacks and a vibrant blue sweater, her blond hair loose around her shoulders, he nearly swallowed his tongue.

Friend, remember? *Friend!*

He was underdressed for church, but he told himself it didn't matter. He was only going to protect Caitlyn.

"I'll go with you," he said curtly. "I'll stay in the back, out of the way."

"You don't have to," she protested, but he held up a hand.

"I have the day off, we can hang out. Maybe watch a movie." He normally went to the gym, meeting up with his partner, Bruce, more often than not, but figured missing a day wouldn't hurt. "Come on, what do you say?"

"I need to get a few things for Snowball after church," she said. "If that's okay with you."

"Sure thing." He held up his keys. "I'll drive."

"Thanks, Devon."

As they headed outside, he kept a keen eye out for anything suspicious. Unfortunately, he didn't know the apartment residents on sight or their cars either. Once they were on the road, he relaxed a bit. Especially since the church wasn't far, he could see the tall steeple from there.

As he passed a small gas station, a black truck pulled out behind them. He waited for several minutes, hoping it would turn off again.

It didn't.

He was about to ask Caitlyn if it looked familiar when the truck abruptly sped up, coming directly toward them. He yanked the wheel hard to the right, going up and over the curb and into a strip mall parking lot. Caitlyn cried out in alarm, but he came to a stop and turned to look. The truck kept going, moving so fast he couldn't get the license plate number.

This was no accident or random event. He was sure the driver was the same guy Caitlyn had seen last night.

And it appeared as if he knew exactly where Caitlyn lived.

CHAPTER THREE

Caitlyn's heart pounded so loudly she was surprised Devon couldn't hear it. When he'd brought the SUV to a jarring stop, she'd twisted in her seat to get a better look at the black truck. It was already speeding away.

"Do you think that was him?" she asked breathlessly.

He nodded. "I won't lie to you, Caitlyn. Who else would come after you?"

Absolutely no one, she thought with a sigh. "I don't understand how he found me so fast."

"I don't know," he admitted.

"Nate probably blabbed." She drew in several deep breaths to calm herself.

Devon's gaze sharpened. "Nate who?"

She waved a hand. "The guy who invited me to Flannery's to see the band. Although why he bothered, I have no clue. He was clearly there to drool over the lead singer."

A muscle flexed along his jaw. "Do you have a last name for Nate?"

"Powers." She glanced at him. If she didn't know better,

she'd think he was jealous. "I think Nate is a regular at Flannery's. I told Detective Ernest about him."

"That's good." His expression turned grim. "Are you sure about going to church? The driver of the truck is gone for now, but he might come back."

Come back? It struck her how close they were to her apartment. As much as she wanted to attend church services, it didn't seem smart to do that now. "Snowball," she whispered. Reaching out, she grasped Devon's arm. "I need to get Snowball."

He nodded. "Okay, we'll go back for the cat. From there, I'll take you to my place."

She was touched by his offer. "Are you sure? I don't want to bring the danger to you."

"I'm sure, and I'd rather protect you than leave you alone at the apartment." He shifted the car into gear and pulled out of the strip mall parking lot.

"Uh, Devon? I still need to get supplies for the cat."

There was a moment's hesitation before he nodded. "Okay, fine. But I'm not sure how early the stores open."

She pulled out her phone and found the pet supply store. "They open at ten."

"By the time we pack up your stuff and get the cat, they'll be open in time for us to stop by on the way to my house." He glanced at her. "But we can't linger. I don't want this guy to find us again."

Yeah, she didn't want that either. She swallowed hard and kept an eye on the side mirror as Devon made his way back to her apartment building.

He chose to park in the closest spot to the door and followed right behind her as she led the way inside. "I'll just be a minute."

"Take your time. The store won't open for ten more minutes anyway."

He waited patiently while she packed her things in her small pink suitcase and gathered the few items she had for Snowball into a bag. Not the makeshift litter box, though, she figured she'd buy a new one for Devon's house. He was already going above and beyond to help her out, the least she could do was try not to make a mess. Lastly, she cradled the cat in her arms, using the same red scarf to keep her warm.

"All set?" He took both the suitcase and the bag from her. "Let's go."

There was no sign of the black pickup truck when they emerged from the apartment building. She had to believe Nate had told the murderer her name, not that Nate had any way of knowing what the guy had done. Still, why hadn't he just kept his big mouth shut?

Devon stored her suitcase and bag in the back, then slid behind the wheel. "If I remember correctly, the pet store isn't far from the clinic."

"Correct," she said, secretly pleased he knew where she worked. Two months ago, he protected her in several locations, including driving her and Annette to school, but never at the clinic. At the time, her schedule had been to only work Friday and Saturday.

"Once we're settled at my place, I need to call Detective Ernest," Devon said. "He needs to know about the black truck that tried to rear-end us."

She nodded. "That should help give credence to my story."

Devon darted her a quick glance. "Did you get the sense he didn't believe you?"

"The boot print helped," she acknowledged. "But he basically said it was my word against his."

Devon scowled but didn't say anything more. He pulled into the pet shop parking lot the moment it opened.

She pushed open her door. "You can wait here, I won't be long."

"I'm coming with you." He looked surprised that she still carried Snowball in the scarf. "You're bringing her inside?"

"They welcome pets," she assured him. Inside, she set Snowball and the scarf in the front of a cart and made her way through the aisles. By the time she had the cat litter, a litter box, cat food, and treats, her total was enough to make Devon's eyes bug out.

"I'll get that," he offered, despite the shocked expression on his face.

"No, she's my responsibility." Caitlyn handed over her credit card.

"That's a lot to pay for a cat that you will only have for a few days," he said as they returned to the SUV.

"I know." There was no denying a tiny part of her wanted to keep Snowball, but of course, she would never take a pet that belonged to someone else.

And it wouldn't be right to pray that the cat didn't have an owner either.

She lightly stroked the kitten's fur as Devon drove to his house. His house was small and cozy, located at the end of a dead-end street. It had a one-car garage, which was an added plus. It looked nice and reminded her a little of the house she and Jayme had lived in for the past few years. Well, before she and Annette had moved into their apartment.

Jayme had offered to either sell the house and split the

proceeds or simply turn the mortgage over to her and Annette at the first of the year. They'd have to break their lease, and they'd have to be able to afford the payments and the insurance.

Annette wasn't too keen on that idea, and Caitlyn couldn't afford the place on her own. As much as she hated the idea, she'd told Jayme to put the house on the market now that she was living with Linc at his home. They were hoping to have a buyer before the holiday.

"It's not big, but it's private, especially being at the end of the road," Devon said, misinterpreting her silence. The automatic garage door opened, and he drove inside.

"It's perfect," she said with a smile. "Honestly, a home like this is so much better than an apartment."

"I was fortunate to get this before the housing market went crazy," he admitted. "There's two bedrooms, but I tend to use one for an office."

"No worries, I can sleep on the sofa."

He hauled everything inside, so she only had to bring Snowball. She was a little disappointed that Devon didn't have any Christmas decorations up but told herself to get over it. The cat meowed as if annoyed to be in yet another unfamiliar surroundings. But once Caitlyn unpacked the litter box and set it up in the farthest corner of the bathroom, the kitten immediately began to explore the area, making herself at home.

Of course, providing more tuna helped.

As she sat in the corner of Devon's sofa, stroking the cat, Caitlyn wondered how long she'd be forced to impose on his kindness. Not that he'd made her feel like she was a pest, but by tomorrow morning, they'd likely both have to go to work.

Would the guy in the black truck find her at the clinic? The thought made her shiver.

She closed her eyes and prayed that Detective Ernest would find and arrest this guy before the end of the day.

DEVON DIDN'T UNDERSTAND why his house seemed smaller now that he had Caitlyn and Snowball staying with him. Maybe he just wasn't used to having guests, having always preferred to live alone.

He went into his room to change his clothes. He also reached for his shoulder harness and weapon, deciding he wouldn't be caught off guard like that again. Feeling better that he was armed, he headed back into the living room. Pulling out his phone, he called the police station, asking to speak with Detective Ernest. He was put on hold for almost five minutes before they were connected.

"Rainer? What's going on?"

"Our suspect knows Caitlyn's identity. A black pickup truck pulled out of a gas station near her apartment as we drove by. He sped up in an attempt to rear-end us, but I drove up and over the curb to avoid the collision."

"We? You and Ms. Weston were in the vehicle together?"

"Yes." Devon glanced over to where Caitlyn sat curled up with her cat. "I ended up staying at her place last night. Based on what happened this morning, it's a good thing I did."

"You're sure it's the same guy?" Ernest asked.

"No, how could I be sure of that?" Devon raked his hand through his hair and began to pace. "Caitlyn saw the guy standing near a dark pickup truck. You don't believe in coincidences any more than I do."

Ernest gave a heavy sigh. "You were in your vehicle, though, right? Not hers?"

"Yeah. But I parked my SUV right next to her Honda. Anyone watching her apartment would have seen the two of us together."

"Where are you now?"

"My place. Did you get the guy's name from anyone at Flannery's?"

"No, the canvass didn't give us any leads. There were several women there with long dark hair, but without more to go on, we haven't been able to pinpoint a potential victim."

He didn't like the sound of that. "No reports of missing persons fitting that description?"

"Nope."

Great. The news wasn't at all reassuring. "I don't get it. How did this guy find Caitlyn so fast?"

"He may have beat us to Flannery's and asked around. Caitlyn is a beautiful woman, someone may have known her name."

Yeah, like Nate Powers, the guy she'd gone to meet. There was no reason for him to be jealous of the guy, it sounded as if the date didn't go well. Especially since Caitlyn had left the place alone. He pulled himself together with an effort. "I can keep her safe today, but tomorrow I have to work. And she probably does too."

"I'm doing my best here," Ernest said, his tone somewhat defensive. "If I get a new lead, I'll let you know."

"Thanks. I appreciate that." Devon knew the detective didn't have to keep a lowly street cop up to date on his investigation, so he was grateful for what little he'd learned. "Take care, Detective."

"You too."

Devon lowered his phone. Caitlyn's expectant gaze made him grimace. "No news, I'm afraid."

Her gaze dropped to the kitten curled in her lap. "Well, there's still time. I'll keep praying that God gives Detective Ernest the skill and wisdom to find the killer."

It was disconcerting, the way she talked so casually about praying. It made him wonder what it was like to have that level of faith in a higher power.

"Okay, so, are you interested in watching a movie?"

"Sure." Her lopsided smile didn't reach her eyes.

"You like Christmas movies?" He tried to remember what he had on DVD. "We can watch *Die Hard*."

She laughed, and this time he was pleased to see her entire face light up. "I can't believe you think *Die Hard* is a Christmas movie!"

"Well, it takes place during Christmas. As a cop, it holds a special place in my heart. And uh, unfortunately, I don't have anything better." He crossed over to sit beside her. Snowball was like a tiny white chaperone between them. Which was a good thing because he liked spending time with Caitlyn, more than he should.

Friends, he reminded himself firmly.

"Uh-oh," Caitlyn said, frowning down at the kitten. "This isn't looking good."

"What isn't?" He bent over to see what she was referring to. Their heads were so close he became distracted by the lavender scent of her hair.

"Snowball's wound." She frowned as she gently palpated the animal's side. "See how red it is? And how it's oozing? I think it's infected."

Poor kitty. "Do you have animal antibiotics?"

"I don't, but the clinic does." She lifted her gaze. "Would you mind taking us there? I have a key, and I can get the antibiotics and leave cash and a note for Dr. John. I know he won't mind, he's always encouraged us to help ourselves to whatever we need as long as we keep track of it. One of the perks of working there."

"Of course." He glanced down at the feline. "But you have to tell me this, how on earth are you going to get her to take antibiotics?"

"It won't be easy," she acknowledged. "It's also a two-man job, so you'll have to help me."

"Me?" He tried not to look horrified. "But I don't know anything about giving cats pills."

She stood, lifting Snowball into her arms. "You'll learn. And really, all you have to do is to hold her."

Yeah, somehow he didn't think it would be that easy. Still, he escorted her and Snowball outside and into his SUV. Devon didn't mind driving her to the vet, but it was possible this guy had his license plate number. After all, he'd figured out that Devon was taking Caitlyn away from her apartment.

He should have had Caitlyn come to his place last night rather than leaving his SUV parked outside next to her Honda. The fact that this guy had found her so quickly really bothered him.

"Keep an eye out for any sign of the black pickup truck," he advised as he backed out of the garage.

She nodded and alternated between looking down at Snowball and watching the traffic through her window. He took the most direct route to the veterinary clinic. Because the place was closed, the parking lot was completely empty. Which only made his blue SUV that much more noticeable.

"This won't take long, right?" he asked as he swept his gaze around the area, searching for anything suspicious. "I'll feel better once we're back at my place."

"No, not too long. Although I would like to get the first dose of antibiotics in her as soon as possible," Caitlyn said, pulling her keys from her purse. "I should have asked you to bring us here earlier. I have no idea how long she's been out in the elements. That laceration could have been festering for days."

"We're here now." He followed her inside the clinic. She had a key, so they weren't breaking and entering, but he still found himself whispering. "You're sure you won't get in trouble for this?"

"I'm sure." She turned and thrust Snowball into his arms. "Hold her for a few minutes, I need to find the antibiotics."

He stared down into Snowball's blue eyes. Thankfully, she didn't seem to be upset with the change in companionship. "She'll be back soon, don't worry."

Meow.

He chose to believe the cat was agreeing with him. Gazing around the clinic, he found himself smiling at how it was brightly decorated for Christmas, including a large fake Christmas tree in the corner filled with all sorts of animal ornaments. The holiday spirit was alive and well here too.

Caitlyn hadn't mentioned his lack of Christmas decorations, but maybe she assumed he hadn't made time to put them up rather than the real fact that he simply had never bothered to decorate for the holiday.

Somehow, he sensed she wouldn't be okay with that plan. Although her opinion shouldn't matter since once this guy was found and arrested, he wouldn't see Caitlyn again.

Except as friends.

"Devon? Bring her into exam room three," Caitlyn called.

He did as she requested, surprised to see that she had several items out on the back counter. "You're not doing surgery on her or anything."

"No, but I want to check her temperature and her heart."

"Okay." He continued holding the cat as she gently inserted an electronic thermometer into the cat's backside. The animal didn't care for it, and he didn't blame her. "It's okay, Snowball. You're going to be fine."

"She's running a slight fever," Caitlyn murmured.

"The antibiotics should help that, right?"

She nodded and then listened to the cat's chest. He knew she was only a tech, but she appeared to know what she was doing. "Okay, we're going to put this soft collar on her to stop her from licking the wound. I'd use a soft jumper, but the area will probably continue to drain a bit. Then we'll give her the antibiotic." She snapped her fingers. "Wait, I should check for a microchip first."

She disappeared through the rear door of the clinic, returning a few minutes later with a wand-like device. She moved the wand slowly over the back of the cat's neck and back. With a frown, she repeated the movement.

"That's weird, no chip," she said.

"Maybe she's a stray."

"She's litter-box trained, so I don't think she's been running around on her own for very long." Caitlyn set the device aside and deftly placed the soft red collar around the animal's neck.

Meow, meow!

Okay, those were definitely not happy meows. Those

were get-this-blasted-thing-off-me-right-now meows. And he couldn't blame her.

"I know, it's awful," Caitlyn soothed. "But your wound needs to heal."

Caitlyn instructed him to hold the cat again while she pried open the cat's mouth. The animal wiggled in an attempt to get away, but he held firm so Caitlyn was able to get the pill down into the animal's throat.

Meow, meow!

"She's going to hate us by the time this is over," he muttered.

"No, she's just feeling lousy because of the infection. She'll perk up in a few days." Caitlyn began to put the supplies away. "Take Snowball out to the lobby. I need to clean up the room, and we'll be ready to go."

"Sure." He carried the cat into the main waiting area, smiling when he saw the note and cash Caitlyn had left on the counter.

When she was finished, he shifted the feline back into her arms. Crossing to the doorway, he peeked out to make sure the black truck wasn't lurking nearby.

No reason for the driver to think they'd be at the veterinary clinic, the guy probably hadn't seen the kitten in Caitlyn's lap as she drove by earlier that morning. But he wasn't about to take any chances either.

Maybe Nate the not-so-great date had mentioned where Caitlyn worked. The guy could right now be scoping the place out, intending to make a move in the morning.

It occurred to him that he might have to take the day off in order to keep an eye on Caitlyn. A bridge he'd cross later.

"See something?" Caitlyn asked from behind him.

"No. It's clear." He opened the door wider and used his fob to unlock the SUV. Then he quickly escorted Caitlyn over to the passenger side door. Once she was safely inside, he jogged around to get in as well.

When he pushed the start button, though, nothing happened. He scowled and tried again.

Still nothing.

A warning shiver slid down his spine. He once again glanced around the area, trying to pick out someone watching them, but from what he could tell, everyone was minding their own business.

He tried one more time, then was about to open his door when he caught a flash off in the distance. "Get down!" He reached over and yanked Caitlyn down just as the windshield shattered into zillions of pieces.

"What's happening?" Caitlyn cried.

"Shooter, roughly fifty yards from here. Call 911 but stay down." He pulled his weapon and reached up to sweep the remnants of glass out of the way. "I'll hold him off."

"But I don't see him," Caitlyn protested.

He didn't either, but the guy was out there somewhere.

She set the scarf-bound cat on the floor beneath her feet and grabbed her phone. He listened as she spoke to the emergency operator, giving their location in succinct sentences. Devon couldn't help but be impressed by how cool she was in an emergency.

Crack!

Another gunshot echoed around them. He wasn't sure if the guy had hit another part of the SUV or not, but since the car had been disabled, they couldn't escape. Or risk running back inside the clinic.

They were trapped!

CHAPTER FOUR

Caitlyn crouched down in her seat as gunfire echoed around them, whispering reassurances to Snowball. Or to herself. Or to both of them.

Never in her life had she been the target of gunfire. Sunday mornings were usually quiet and tranquil in Sevierville. Especially during the holiday season.

"Caitlyn? Are you okay?" Devon's voice was tense. Glancing up at him, she saw a combination of fear and anger etched into his features.

"I'm not hurt." She wasn't okay, not by a long shot, but she understood he was concerned about her safety. Glancing down at her phone, she realized the 911 operator was still on the line. She pushed the speaker button. "He's still shooting at us, are you sending help?"

"Yes, squads are on the way. Is anyone hurt?" the operator asked.

"Not yet," Devon said. "But we will be if you don't get someone here right away!" Another gunshot punctuated his statement.

"Hurry," Caitlyn urged, but seconds later, she heard the

wail of sirens growing louder and louder as several cops responded to the scene. She disconnected from the call and peered up at Devon. "Will they search for him?"

"Yes." He lifted his head to look out through the window, so she did the same. It appeared as if the entire police department had arrived, the officers all wearing full body armor, including helmets, as they used their squads to surround their badly damaged SUV.

"Shooter was somewhere to the west," Devon shouted, "using a rifle with a scope."

Several officers broke away from the group to head west. Two other officers came over to open Devon's driver's side door, staying low and using the open door as a shield. "Any injuries?"

"No, we're okay." Devon glanced over, and she nodded in agreement. "I think the shooter sabotaged my car, making it impossible for us to escape."

The officers exchanged a grim look. "How did the perp know you would be here?"

Devon shook his head. "I don't know. I don't think we were followed to my place. If we had been, why wouldn't he have come after us right away? We would have been easier targets there."

His words made her shiver because he was right. Just imagining the shooter finding them at Devon's home made her feel sick to her stomach. Yet she couldn't figure out how the guy had managed to find them here at the clinic. Had he been hanging around the area, waiting for them to come out of hiding?

Or had he been hiding nearby, waiting for her to show up at work the following morning?

Neither scenario was the least bit reassuring.

"Let's get you two out of here," the officer said. "We'll take you one at a time over to the squad."

"Take Caitlyn first, Bobby," Devon said. "She's the target."

"Okay, Caitlyn first," Bobby agreed. "But by the looks of your SUV, I don't think she's the only target. Looks like this guy didn't care about taking you out along with her."

Devon shrugged this off as if it didn't matter, but Caitlyn hated knowing she'd dragged Devon into danger. All because she'd called him in the middle of the night.

Devon slid out of the seat, crouching next to the two officers. He gestured for her to crawl over the center console to join them. She darted a glance through what was left of the shattered windshield, then gathered Snowball off the floor and set her in Devon's lap. "Take her, will you?"

"A cat?" Bobby frowned.

"I'm not leaving her here." Caitlyn didn't care what the cops thought, she wasn't going anywhere without Snowball and the supplies she'd taken from the clinic. She grabbed the bag and then quickly scooted over the center console to join Devon.

Thankfully, Snowball liked Devon. She sat nestled in the red scarf as he held her against his chest as if she belonged there.

At this rate, she wouldn't be finding Snowball's rightful owner anytime soon.

She tucked the bag under her arm and reached for the cat, but Devon shook his head. "I'll bring her. Go with Bobby, okay?"

For a moment she hesitated, then nodded. Devon wouldn't leave the cat behind. The two officers stood and kept her sandwiched between them as they escorted her to the car. They were tall, so she couldn't see if the other cops had found the shooter.

Somehow, she didn't think they had.

She slid into the back of a squad and waited for Devon. Moments later, he joined her, gently setting the cat in her lap.

"Thanks," she murmured, stroking Snowball's fur.

"We'll need to go to the police station to give our statements," Devon said. "Then we'll need to find a new place to stay."

"Like where?" She glanced down at the cat. "We'll need to pick up all the stuff I bought earlier."

"I know." Devon's gaze softened. "Bobby offered to go to my place and get what you need."

"That's nice of him." She frowned. "You really think it's not safe to go to your house?"

"I won't take a chance with your life, Caitlyn." His gray eyes were somber. "This was a premeditated attack. First going to the trouble of disabling our vehicle, then shooting at us from a safe distance." He sighed and rubbed the back of his neck. "Since I have no idea how we were found, we need to go someplace different."

"Like the cabin?" she asked, referring to the log cabin she and her sister had used as a safe house two months ago. Then again, the bad guy had found them there because of her. Then she considered another option. "We could go to Jayme's house. I mean, it's listed with a realtor, but it's empty."

"That's a potential link to you too. No, I'm leaning toward a motel rather than the cabin. The cabin didn't work so well last time," he said grimly.

"Whatever you think is best." She wasn't about to argue.

Devon was a cop, a good cop. He was an expert at this type of thing.

She closed her eyes, wishing she'd never witnessed the man strangling the woman in the woods. Then a feeling of shame washed over her. If she hadn't witnessed the crime, the woman's family would never know what happened to her. They'd never get the justice they deserved.

No, God had guided her to that location earlier that morning. This was the path He'd chosen for her. She didn't know why, but it wasn't up to her to question God's will. All she could do was trust in Him.

And in Devon's ability to keep them safe.

"Any sign of the shooter?" Devon asked as Bobby slid behind the wheel.

"Negative." The cop glanced over his shoulder at them. "I'm sure he's long gone."

"Yeah." Devon sighed heavily. "I was just hoping he'd slip up."

"He will, they all do." Bobby started the car and drove out of the parking lot. She felt vulnerable being out in the open like this. The smallest sound made her flinch.

"Hey, we're safe," Devon said, reaching over to take her hand in his.

"I know." Easy to say, but controlling her reactions was impossible. Even escaping the horror of the Preacher hadn't caused this level of fear. Mainly, she knew, because her older sister Jayme had shielded her from the worst of things.

Now she was the center of some murderer's attention. Someone who wouldn't stop until she'd been silenced, once and for all.

Deep down, she couldn't help but wonder if she would ever feel safe again.

DEVON COULDN'T BLAME Caitlyn for being on edge. In the span of twelve hours, she'd witnessed a brutal murder and had been targeted twice. The attempt to rear-end them was nothing compared to being shot at in a disabled vehicle.

Clearly, he'd underestimated this guy. The killer was smart and cunning, and very determined to get to Caitlyn.

A steely resolve had him clenching his jaw. No way was he going to allow this guy to hurt her. The scumbag would have to go through Devon first.

Unfortunately, the assault on them outside the clinic indicated the guy had no problem taking down whoever stood in his way.

Including a cop.

Upon arriving at the police station, he and Bobby kept Caitlyn between them as they went inside. She carried Snowball, and he hoped the cat would be able to hold on a little longer. The police station didn't have a litter box or any cat food.

"You know I need to interview the two of you separately," Bobby said.

"Yeah. Do me a favor and take Caitlyn's statement first." He wanted to get in touch with Detective Ernest for an update on the investigation.

"Why not?" Bobby turned to Caitlyn. "This way, please."

After they were settled in one of the interview rooms, Devon prowled around the empty police station. He wasn't surprised no one was around, they'd all gone out to the scene of the shooting. That kind of thing wasn't a usual occurrence in Sevierville.

Until today.

He'd half expected to find Detective Ernest there, but maybe he'd gone to the clinic too. He quickly called him.

"I know about the shooting," Ernest said in lieu of a greeting. "I'm here now."

"This guy is escalating." Devon knew he wasn't telling Ernest anything he didn't already know. "Strangling his girl-friend may have been a crime of passion, but this was straight up premediated murder."

"I get it," Ernest said, his tone testy. "I'm following every lead."

Devon blew out a frustrated breath. No reason to take his anger out on the guy who was doing his best to find the perp. "I know you are. I can't believe the canvass at Flannery's didn't give you any information on this guy."

"Travis Karp, the bartender, remembers him standing in the corner, exactly the way Ms. Weston told us. But he drank whiskey neat and paid in cash. According to the bartender, the guy isn't a regular, at least not on Friday or Saturday nights when Travis normally works. So far no one else seems to know his name either."

Devon frowned, trying to understand why this guy would have been there and what would have upset him to the point where he'd strangled a woman to death? It didn't make much sense. "No missing person reports yet either?"

"Not since the last time we spoke, what, ninety minutes ago?"

"Okay, okay." Devon could tell Ernest was getting annoyed with him. "After that near miss outside the veterinary clinic, I had to ask."

"Look, I understand how rough this is for you," Ernest admitted. "I'd be just as angry about someone coming after me and someone I cared about."

"Caitlyn is just a friend," he hastily interjected.

Ernest ignored him. "It's personal for you, Devon. You need to back off and let me do my job. I promise to keep you in the loop on any new leads."

"Thanks, Wade. I appreciate it." Devon disconnected from the call and glanced around the empty police station. His being a potential victim would make every cop on the force that much more determined to find this guy.

The brothers and sisters in blue stood together in times of crisis.

He thought about calling Bruce as backup, then decided against it. His partner deserved a couple of days off, they'd been short-staffed and had worked a lot of extra hours recently. Bruce and his wife needed some time together.

Devon thought about the unknown victim. The murder had only taken place less than twelve hours ago, but he still found it odd that no one had reported her missing. And it was strange that Caitlyn hadn't seen the victim at Flannery's. It made sense both the man and the woman had been there together at some point, otherwise how had they ended up in the woods at the same time?

Way too many questions without a single, viable answer.

He continued pacing, his mind whirling. How had the shooter found them? It wasn't as if they'd planned a trip to the veterinary clinic.

But Caitlyn worked there.

The knowledge hit hard. The shooter would have been waiting for her to show up on Monday morning. Maybe had even scouted the place out today. He must have been watching as they'd pulled into the parking lot and decided to make another attempt at them.

He hoped this scenario meant the guy hadn't registered his license plate during the attempt to rear-end them.

Although he couldn't discount the possibility either. Devon's car would be towed to the police garage and examined for evidence, but that made him realize he'd need another vehicle.

Caitlyn's Honda wasn't an option.

There were unmarked vehicles available. Normally used for undercover types of operations and only with the captain's approval. Devon figured this current situation met that criteria hands down.

When the interview room door opened, he had a semblance of a plan. First, he'd have to give Bobby his statement, then the officer would need to pick up all the cat paraphernalia at his place. By noon, he hoped to have Caitlyn settled in a new and safer location.

After explaining the events of both the shooting at the clinic and the earlier near collision, Bobby tapped his pencil and asked, "Do you believe Caitlyn's claims that she has no idea who this guy is?"

"Yes, I believe her." Devon frowned. "She has no reason to lie, Bobby. I'm telling you, she looked scared to death when she explained how that guy had strangled the woman. And she wants him caught, just as much as we do."

Bobby lifted a hand. "Okay, but I had to ask. This kind of thing doesn't happen here."

Devon knew Bobby was right. However, it was just a few short months ago when an arsonist had targeted Caitlyn's sister Jayme. So yeah, unfortunately this kind of thing did happen.

He dug out his keys and slid them across the table. "I need you to pick up all the cat stuff from my place and bring it here. I'm going to check out one of the unmarked vehicles."

"Sure." Bobby picked up the keys. "She's really attached to that thing."

Devon didn't want to admit how much he'd become attached to the feline. He knew better than to get too attached since the cat probably belonged to someone else. Caitlyn would be sad once she had to turn Snowball over to her rightful owner too.

But finding Snowball's owner wasn't the top priority. As Bobby left with his keys, Devon found Caitlyn sitting near one of the workstations.

"Did the interview go okay?" He dropped to his knees beside her. He stroked Snowball's fur, the cat seemingly content to be snuggled up in Caitlyn's red scarf.

"Yes." Her expression was grave. "I told him everything I know. But it seems as if they're no closer to finding this guy."

They weren't, but he tried to smile reassuringly. "They will. I spoke with Detective Ernest. The bartender at Flannery's remembered seeing a tall, dark-haired guy standing at the rear corner of the bar, but he didn't recognize him as being one of their regulars. And he paid in cash, so no way to trace him through a credit or debit card."

She perked up at that news. "I'm glad I wasn't the only one to notice him."

"You weren't." He gave Snowball another stroke, then stood. "Stay put for a little longer. Bobby went to my place to get Snowball's litter box and food. In the meantime, I'm going to find us a new vehicle to use." He paused, then asked, "Have you spoken to Jayme?"

"No, and I'd rather not interrupt her and Linc on their honeymoon."

"I understand you want them to have fun, but this is a

big deal, Caitlyn. Linc will be upset if he finds out we didn't bother to loop him in."

"Please don't call them," Caitlyn begged. "They'll feel obligated to come home, and I don't want to ruin their trip."

"Caitlyn." He sighed. "I don't feel good about keeping this quiet."

"Just give it a few days." She reached up to take his hand. "I feel certain that detective of yours will find the man responsible."

It was hard to resist her pleading gaze. "I'll give it another twenty-four hours," he finally said. "But if there's any more danger, then I'll need to let Linc know."

"Thanks, Devon." She squeezed his hand. "I know I'm safe with you."

Her trust in him was humbling. He desperately hoped he wouldn't disappoint her. He wanted nothing more than to pull her into his arms, but he forced himself to turn away.

Ernest was right, this was personal. More so because of his growing feelings for Caitlyn. Linc would not be happy to know how much he wanted to kiss his wife's younger sister. Especially since Devon had been a bit of a playboy the past few years since Sabrina had left him. Devon admired Linc, he was a good investigator and had proven to be an excellent detective, figuring out where Jayme was being held by her kidnapper.

In fact, Linc was the one who'd encouraged Devon to sign up for the detective exam, which he was scheduled to take after the first of the year.

No, he couldn't pay Linc back by crossing the line with Caitlyn.

He was already regretting his promise to wait twenty-four hours to call Linc and Jayme.

Devon crossed over to the closest computer and logged

in. Their boss kept an inventory of vehicles, and he scrolled through them until he found another SUV. He typed in his name as the person taking accountability for the vehicle, then shot off a message to his boss, explaining the need to have it. When that was done, he went over to the cabinet to get the keys.

Their small fleet of vehicles was parked out back. He went outside to make sure it was there and filled with gas. Satisfied, he returned to where Caitlyn patiently waited.

"We're all set." He pulled a chair over to sit beside her. "As soon as Bobby gets here, we'll go."

"Go where?" Caitlyn lifted anguished eyes to his. "A nice hotel isn't going to want me to bring a cat."

She had a point. He figured anyone hearing Snowball meow might turn them in. He used his phone to search for pet-friendly hotels. "Okay, I found one. Only three stars, but they allow pets for an additional fee." He turned to show her the phone.

"The Mountain View Motel." She nodded. "I've never been there, but it looks okay. I've been in far worse places, that's for sure."

He frowned. "Far worse? Like where?"

"Jayme and I stayed in the Shady Lane for several months." She wrinkled her nose. "I despised it."

Every cop on the force knew about the Shady Lane, crimes of all types took place there on a regular basis, especially drug dealing and prostitution. He tried not to look horrified. "I'm sure it was awful."

"It was. But at the time, we were fortunate to have a roof over our heads." She shrugged as if living in a terrible place like the Shady Lane was no big deal. "Living in the woods during winter was worse."

He'd always known Caitlyn and Jayme had struggled

early on, but this level of detail was a shocker. Oh, he still believed she was innocent as far as relationships went, but Caitlyn was portraying a level of maturity that you didn't often see in twenty-three-year-olds.

Maybe he'd underestimated her. Before he could say anything more, though, Bobby called his cell. "Find everything?"

"Yeah, I think so." There was a pause, then his fellow officer added, "I also think someone has been here."

Devon leapt to his feet. "What makes you say that?"

"I could be wrong, but it looks as if the house has been searched. I see kitchen drawers that are not closed all the way, things moved around. After everything that's happened, it's all a bit suspicious."

"How do you know stuff has been moved?" A trickle of unease skated down his spine. He and Bobby worked together, had maybe hung out on occasion, but the cop had never been to his house. Granted, he'd been in plenty of situations where he'd trusted Bobby to cover his six. To back him up when things got heated.

Had that trust been misplaced?

"Dude, you're not exactly a star housekeeper," Bobby said. "There's plenty of dust on your end tables. Looks as if the dust has been recently disturbed, especially near the stack of books."

"Were the doors locked when you go there?" He wasn't sure how his house had been accessed.

"The front and back doors were, yes. But not the one connecting the house and the garage."

"I rarely lock that one." Something he'd absolutely do from now on. "I don't know how anyone would have gotten my key code for the garage door, though."

"Yeah, that's how I assumed the intruder got in."

Caitlyn was listening to his side of the conversation. He put the phone on speaker and set it on the closest desk. "Caitlyn is here with me." Devon looked at her. "Did you move stuff around on my end tables earlier today? Maybe looking for something to read?"

Her blue eyes widened. "No. I didn't touch anything other than the items I bought for Snowball. Why?"

"Listen, I could be wrong," Bobby said. "It's a sloppy job of searching if you ask me. And as I said, no sign of forced entry. But I wanted to double-check with you while I was still here."

"I appreciate that." Devon didn't think Bobby was over-reacting. But who would bother to search his house? The shooter? When? Must have been after he'd escaped from shooting at their SUV outside the veterinary clinic.

Which meant their suspect knew where Devon lived.

Not good.

Thankfully, he'd already made the decision not to take Caitlyn back there. Still, the way this guy continued to be a step ahead of the investigation was starting to tick him off.

He needed to figure out who this guy was, and soon. Before Devon failed in his mission to keep Caitlyn safe.

CHAPTER FIVE

Caitlyn was horrified that Devon's house had been searched, but even worse was the fact that the killer knew who Devon was. And that Devon was protecting her.

Goosebumps rose on her arms as she looked around the nearly empty police station. Had the guy known they'd end up here after his failed attempt to kill them outside the veterinary clinic?

Was he outside right now waiting and ready to strike again?

"Don't worry, I'll get you safely out of here," Devon said, breaking the silence. "As soon as Bobby gets here with the stuff from my place, we'll find a place to stay."

"I know." She met his gaze. "But I feel terrible that this guy knows who you are and where you live."

"I've been wondering about that," Devon admitted. "I don't like the thought that he was watching your place. He obviously could have gotten my license plate number while my vehicle was parked next to yours outside the apartment. Or again when he tried to rear-end us. But how he'd traced the SUV to me, uncovering my name and address, is concerning."

She shared that concern. More so because she was the one who'd dragged him into this mess. She should have just called 911 rather than asking Devon to come meet her at the apartment.

Although if she hadn't called Devon, it was likely the murderer would have succeeded in finding and killing her too.

Why, God? Why?

Going to church, learning about God's word in the Bible had brought her a sense of peace. Here she was, once again, wishing she understood God's plan. Why He was taking her and Devon down this path.

"There's Bobby." Devon's voice cut through her thoughts.

"Great." She tried to smile despite the anxiety churning in her gut. The young cop hauled a large bag of items while rolling her pink suitcase through the myriad of desks and cubicles to join them.

"I hope I got everything," he said, setting the bag at her feet. "I figured this was yours and not Devon's."

"It is, yes." Keeping Snowball tucked in her lap, she bent over and rifled through the contents. In addition to the cat items, she noticed he'd packed some clothes for Devon. "You did great. Thanks so much."

"You're welcome." Bobby shifted to face Devon. "Did you get a vehicle?"

"Yeah, I grabbed the dark gray SUV." Devon reached for the bag of cat items and her suitcase. "Time for us to get out of here."

"Where are you headed?" Bobby asked. "Do you want me to follow you?"

She was about to agree to his following them, but Devon shook his head. "Nothing personal, Bobby. You've been a huge help, but at this point, I'd rather no one knows where we're going. I hope you understand."

"I get it." Thankfully, Bobby didn't take offense. "Just let me know if you need something. This guy isn't your average killer."

"I will, thanks." Devon glanced at her. "Ready?"

She stood, cradling Snowball to her chest. The cat had done remarkably well throughout the morning, yet she needed to get the litter box set up sooner rather than later. A kitten's bladder wasn't very big.

Devon led the way to the back door of the police station. "Wait here. I'll put this stuff away first, then come back for you."

Five minutes later, they were on the road. She noticed Devon kept a keen eye on his rearview mirror. His hypervigilance was appreciated, she was alive right now because of Devon's cop instincts. But she still wished she hadn't pulled him into her problems.

The haunting memory of the woman being choked to death flashed in her mind. She blinked and pushed it away.

Finding the murderer was the key to getting justice for that poor woman.

"I thought we were going to the Mountain View Motel?" She noticed Devon was headed in the opposite direction. "It's to the south, isn't it?"

"Yes, but I'm taking extra precautions." His features were grim. "I don't want any nasty surprises to show up later."

"I hate to ask, but what about tomorrow?" She stroked Snowball's fur. "I work at nine."

"You can't go to the clinic, Caitlyn." His tone was firm. "Not until this guy is found and arrested."

"It's my first full week of work." She understood his reasoning, but she didn't want to let this interfere with her new job. "Dr. John stood by me and Annette while we were in school. Kept us working part-time. With Annette gone in Florida, they need me."

"It was barely two hours ago that we were victims of a shooter outside the clinic," he said sharply. "You can't go in. End of story."

She looked away, staring out the passenger side window. That wasn't the end of the story as far as she was concerned. Maybe she was being pigheaded, but she figured the shooter would expect her to go into hiding, not to show up at work.

After a few moments of silence, Devon glanced at her. "I'm sure your boss won't want anyone to be in harm's way."

The words hit hard because he was right. As much as Dr. John would want her help, she highly doubted he'd want her to bring trouble to the clinic's front door.

She sighed and rubbed her temple. "I'll talk to him. See what he thinks."

Devon nodded and let the subject drop. Her stomach rumbled with hunger; it seemed as if the omelets she'd cooked for breakfast had been days ago instead of hours.

"We'll grab something on the way," he said, turning right, then right again to head back in the direction where the Mountain View Motel was located.

"Something quick and easy because the way Snowball is squirming, I think she needs the litter box."

"What? Don't let her do that in the car." Devon stepped on the gas. "We'll be there soon. Just have her hang on a little longer."

"Sure, I'll tell her that." Caitlyn shook her head wryly. "I'm sure she'll understand how important it is not to pee in the car."

Devon muttered something she couldn't hear and kept driving. It didn't take long for them to reach the motel. "Stay here, I'll be right back." Devon leaped from the car and headed inside. A few minutes later, he returned with two motel keys. "We have connecting rooms at the far end of the row."

"Sounds good." She was touched by his sweet gesture, obtaining two rooms in order to give her some privacy.

He drove to the end of the motel. He unlocked the room closest to them and carried in the cat supplies. She thrust Snowball into his arms so she could set up the litter box. It made her smile that he spoke softly to the cat, letting her know it would be time to go to the bathroom very soon.

"There, set her down." Caitlyn stepped back from the litter box she'd tucked into the corner of the room, as far from the main doorway as possible. She watched as Snowball daintily stepped in and squatted. "Good kitty."

"Well, thank goodness for that. I'll get your suitcase." Devon disappeared back outside.

When he didn't return after several minutes, she cautiously opened the door and peered out. He was smearing mud over the rear license plate of the SUV.

"Isn't that against the law?" she asked.

"Yeah, but I don't care." He rose and stood back to survey his handiwork. "Getting pulled over is the least of our concerns."

No argument there. She turned away, but Devon quickly joined her, rolling her suitcase behind him.

"Thanks." She took the case, feeling a little foolish that she'd bought the bright pink suitcase in the first place.

"Please open your side of the connecting door." He smiled, then went back outside to access the room adjacent to hers. She unlocked the door and opened it as he did the same. "Thanks. I can order lunch and run over to the fast-food restaurant to pick it up. What would you like?"

"A grilled chicken sandwich." She reached for her purse, but he held up his hand.

"This is on me."

"You paid for the rooms, which is already going above and beyond." She was low on funds at the moment, seeing as she'd only been working part-time while attending classes. She and Annette had managed to pay rent, utilities, and food each month, but there hadn't been much left over to put into her savings. But her lack of cash wasn't Devon's fault. She needed to pay her own way.

"My treat," he insisted. "It's the least I can do. Fries?"

She wavered, on both letting him pay and the fries, before nodding. "Sure, fries sound good. But nothing to drink. I prefer water."

"Me too." He flashed a smile. The warmth in his gaze made her all too aware of how attractive he was. Even when she was irritated with him, he knocked her off balance without even trying.

Maybe it was a good thing he'd turned down her offer to have coffee. She wasn't very good at this dating gig. The way Nate had pretty much ignored her last night was proof of that. And her previous date with Duncan had been just as much of a dud. He'd talked about himself, without asking anything about her or her job. Both times, she'd been more than happy to leave.

Nate had outright called her boring when she'd refused to have an alcoholic drink.

His loss, not hers. Although she wouldn't mind

smacking him if it turned out he was the one who'd blabbed about her to the murderer.

Wait a minute. She jumped to her feet. Why not find Nate? At the very least, they'd be able to confirm if he was the one who told the dark-haired guy her name and address. Maybe Nate even knew the guy. Having a name would certainly go a long way in tracking him down.

When Devon returned with food that smelled delicious, she grabbed his arm. "We need to find Nate Powers."

He lifted a brow. "Why?"

She gave him an exasperated look. "Because he was likely the one who gave the murderer my name and address. What if he actually knows the guy?"

"Sit down and eat before your food gets cold." Devon dropped into a chair next to the small table.

"But Nate . . ."

"Detective Ernest is looking for Nate, maybe has even found him by now." He regarded her steadily. "He's working the case, Caitlyn. My only job is to keep you safe."

Dejected, she sat across from him. "I just want my life to go back to normal."

Devon winced and reached for her hand. "I'm sorry, but you must know your life will be forever changed by what happened last night. Even once we find and arrest him, you'll have to ID him, then testify against him. None of that is normal, Caitlyn."

She stared at their clasped hands for a long moment. "You're right. Watching him kill that poor woman has changed me forever. I'll always see that terrified look on her face, hear the pleading in her voice."

Devon rose and tugged her close, wrapping his arm around her. She hugged him tightly, tears pricking her eyes.

She desperately wished she could go back in time to find a way to save that woman's life.

"Shh, don't cry."

Ironically, his words only made her cry harder. After long moments, she managed to get herself under control. "I'm sorry to fall apart on you." She sniffled and took a step back.

"Caitlyn, don't ever apologize for being human." He lifted her chin with his index finger and then slowly lowered his mouth to hers.

A voice in the back of her mind warned her this was trouble, but she ignored it. Instead, she moved closer, reaching up to meet him halfway. The moment his lips captured hers, a sizzling awareness swept over her.

This. The instant chemistry that flashed between them was what had been missing in her previous dates. Making Caitlyn wish she could stay in Devon's embrace forever.

HE'D SUSPECTED KISSING Caitlyn would open Pandora's box, but he'd grossly underestimated the intensity of desire that flared between them. She felt perfect in his arms, and it took every ounce of self-control he possessed to break off from their embrace.

"Uh, um, our food." He did his best to gather his scattered thoughts. "It's getting cold."

Cold food would be welcome at the moment because the motel room's heat level mirrored that of the center of a volcano.

"Okay." He was grateful for her simple response, but when she sat and reached for his hand, a jolt of electricity zinged up his arm. "Dear Lord, we thank You for keeping us safe today. We ask for Your continued guidance as the police seek the man responsible. We also thank You for providing us this safe place to stay and the food we are about to eat. Amen."

"Amen." The response came automatically, despite the fact that he hadn't prayed over a meal since he was a little kid, and that was only at Thanksgiving and Christmas.

"Thanks, Devon." Her smile made a mockery of his willpower. He had to force himself to look away.

Burger. Fries. His hunger for food had vanished beneath the impact of Caitlyn's kiss, but he picked up a fry and ate it anyway.

He needed to stay focused on keeping her safe. Not holding her. And not kissing her or wishing he could do that all again.

Watching Linc get emotionally involved with Jayme while trying to keep her safe had shown him the importance of keeping a professional distance from someone you're tasked with protecting.

He didn't want to make the same mistake again. Especially not with Linc's sister-in-law. Jayme and Caitlyn weren't sisters by blood, but that didn't matter. They had a tighter bond than most family members.

Not that he was an expert on families, having grown up as a pawn for his parents to fight over. He'd never felt comfortable living with either of them, as he'd bounced back and forth as ordered by the court who'd granted his parents joint custody.

Lack of family or what it meant to be a part of a family was another reason to leave Caitlyn alone.

"What are we going to do all afternoon?"

Caitlyn's innocent question made him choke on his

burger. His mind had plenty of ideas, none of them appropriate.

Pull yourself together, Rainer!

"Bobby included my laptop computer," he said after taking a sip of water. "I'm going to search for women who have been reported missing."

"That's a great idea." She beamed at him. "I'd like to help."

"No, thanks." He winced as the refusal came out harsher than he intended. "Actually, you can help, but only after I have a pool of women gathered together. I'm hoping you'll recognize the victim."

"A pool of women?" Her tone was skeptical. "Based on what criteria exactly?"

He ate a french fry to avoid answering. "Well, for one thing, it's probably not someone who was reported missing a year or more ago."

"You don't know that," she protested. "Originally, she may have willingly gone with him, only to discover he became abusive over time. And you can't limit your search to women with a certain hair color either as that's easily changed."

He hated to admit she was right. "I have to start somewhere. I figure looking back the past three months would narrow the victim list."

She shrugged. "I agree that going back three months works as a starting point, but once we get through that group, we should keep going back in time, just in case."

"We?" He frowned. "This is police work, Caitlyn. I'll be accessing my work database, that's not something I can share with a civilian."

"I'm the only witness to the murder." She stared at him.

"I think that moves me out of the usual civilian status, don't you?"

He blew out a breath, sensing further arguing was useless. Besides, she was the only one who could identify the victim. "Fine. You can help."

"Good." She continued eating.

As they finished their meal, he noticed Snowball was playing with a soft stuffed mouse. The cat batted it with her paws, sending it skipping along where she then pounced on it.

Devon could relate. He felt a bit like something being batted around too.

Caitlyn cleared off the table while he unpacked his laptop. The battery was fully charged, so he opened it and logged in. When Caitlyn sat beside him, her lavender scent made him lose focus for a moment.

Oh yeah. The database.

He turned the computer so that he could type in his username and password. Then he shifted the device so she could see it too.

"I'll start with the past thirty days," he said, typing in the dates in the appropriate fields.

"This is a police database across the entire US?" She glanced in surprise.

"Yeah, it's actually funded by the FBI to assist with uncovering victims of human trafficking."

"I had no idea this existed." She frowned. "Apparently, no one bothered to include the seven of us foster kids in there."

"They should have." He frowned and typed in her name. Nothing came up. "I don't understand why you wouldn't have been in the database."

She waved a hand. "Doesn't matter now. Besides, even

if we had been found years ago, we would have escaped again. Honestly, we were better off on our own."

Her matter-of-fact tone made him feel bad for whining about his childhood. Obviously, he had nothing to complain about. Being bounced like a ball between parents was better than living on the streets or in the Shady Lane Motel.

He focused his gaze on the screen as the first image appeared.

"That's not the woman in the woods," Caitlyn said without hesitation.

"Okay." He clicked on the next woman.

"Not her either."

Devon didn't want to doubt Caitlyn's ability to recognize the victim, but her easy dismissal worried him. "Take your time," he advised.

"Her cheekbones were high, and her chin was pointy." Caitlyn gestured to the image. "This woman has a rounded chin."

Okay, then. She obviously had noticed the victim's unique facial features. Well, as much as she could see while the poor woman was being strangled to death.

Caitlyn rejected ten pictures in a row but then frowned as she peered at the next image more closely.

"The hair texture isn't right, the victim's hair wasn't this full." She continued staring at the woman's face for a long moment. "I don't think it's her."

"Maybe we should make a list of possible victims," he suggested. "Maybe there's a reason her hair doesn't look the same as it does in this picture."

"Good idea. This one is close, but I honestly don't think it's her. But it's better to be safe than sorry."

He pulled up a blank spreadsheet and made several columns: name, date of birth, and date reported missing. He typed in Sharlee Corby and her corresponding information. She was only twenty-two, a year younger than Caitlyn.

It made him sick to think about where this young woman might be. Where all these missing women might be. Dead, alive, abused . . .

"I know. It's sad, isn't it?" Caitlyn murmured.

"Yeah." He cleared his throat and hit the keys to bring up the next photo.

They worked for nearly an hour straight before she rose and stretched. "I'm not used to sitting for so long."

"I understand." He sensed that going through these photos was taking an emotional toll. And as much as he'd wanted to protect her, she'd been right to insist on being involved in the search.

Caitlyn was their best chance of finding this guy.

His phone rang, and his pulse jumped when he recognized Detective Ernest's number. "Rainer."

"Devon? Where are you?"

He hesitated. "Someplace safe. Why?"

"I found Nate Powers." Ernest's flat tone gave him a sinking feeling.

"Good, I'm glad to hear it." He lowered the phone and put the call on speaker so Caitlyn could hear. "Does he know our perp? Did he talk to him?"

"I believe Powers knew the guy." Ernest sighed. "But we'll never know for sure because he's dead."

"Wait, what? Powers is dead? How?"

Caitlyn gasped and clapped a hand over her mouth. Her eyes were wide as Ernest continued speaking.

"We found him at home, with a bullet lodged in the back of his head."

CHAPTER SIX

Caitlyn sank onto the edge of the bed, swallowing against a surge of nausea. Nate was dead. She hadn't intended to see him again, but this? Nate didn't deserve to die.

To be murdered. Cold-bloodedly shot in the back of the head.

Snowball meowed and jumped into her lap. Caitlyn cuddled the cat close, grappling with what had happened.

"You think the shooter killed him to cover his tracks?" Devon asked. "I'm sure Powers must have given the guy Caitlyn's name."

"It's one theory," Ernest said. "But I can't rule out the possibility these crimes aren't connected."

"Come on, Wade. You know this is way too much of a coincidence."

"It's part of the job to keep an open mind, Rainer. Something you'll have to learn if you intend to get your own gold shield."

His last comment had her glancing at Devon in surprise. She hadn't realized he wanted to become a detective. In her humble opinion, he'd make a great one.

"I'm aware you need an open mind, but things are escalating. First he strangled a woman, then he tried to run us off the road, then he took shots at us outside the clinic, now this?" Devon raked his hand through his hair. "I thought you were heading over to talk to Powers earlier?"

"I did, he didn't answer the door." The detective's tone held regret. "He wasn't a suspect, so I couldn't just break his door down."

"No, I guess not." Devon shook his head. "Any idea on time of death? Maybe the killer was already there when you came knocking? That made him panic and kill him."

"I'm waiting for the ME to learn more, but I believe he was killed recently," Ernest said. "Like you said, he may have been killed within the past few hours."

Hours. If they'd found Nate Powers earlier, he might still be alive.

And the killer would be in custody.

She understood God had a plan, but at times like this, it was hard to understand.

"Keep me posted once the ME gets there," Devon said. "If our killer did shoot Powers, then he's trying to cover his tracks. Doubling down so that we can't find him."

"I know. Keep a close eye on our witness."

Devon's gaze clashed with hers. The determined expression in his eyes was reassuring. "Trust me, that's the plan."

"I'll be in touch." The detective disconnected from the call.

"I can't believe Nate is dead." She still couldn't wrap her mind around it.

"I know." Devon dropped down beside her, draping his arm over her shoulders. "I'm sorry. I wish we'd have gotten to him sooner."

"Me too." She put a hand over her stomach, willing her lunch to stay put. "It's so awful to know this man kills people at the drop of a hat."

"Two different types of crimes, though," Devon said. "Strangling someone is usually a crime of passion or anger. But shooting a man in the back of the head is an execution."

She paled.

"Sorry." Devon gave her a brief hug. "I shouldn't talk to you as a fellow cop."

"It's okay." She tried to pull herself together. "I'm sure it helps to look at things critically, from a distance."

"It does, but I'm not going to let him touch you." Devon kissed her temple. "You see now that you can't go to work tomorrow, right?"

"Right." After hearing about Nate Powers's murder, it was clear this maniac was capable of anything.

He wouldn't hesitate to kill her too. Or anyone else at the veterinary clinic who got in his way.

The very thought of searching through the missing women on the database made her want to cry. So many women. It hurt to think about what they were going through.

Or that some of them may also be dead.

She resolutely pushed her fears aside and turned toward Devon. "It's important that we get back to work. If our victim is in that database, we need to find her."

"Are you sure you're up to it?" Devon's clear gray eyes seemed to look deep into her soul. "I know it's not easy."

"It's not, but we need to find this guy." She forced herself to set Snowball down and to stand, moving away from Devon's friendly embrace. Leaning on him for strength was fine, but she needed to do her part as the only witness to a brutal murder. She took a seat in front of his computer. "And learning the victim's identity is our best opportunity to bring her killer to justice."

"It is." Devon moved over to sit beside her. "You're a strong woman, Caitlyn."

"I'm trying to lean on God's strength." She darted a glance at his handsome profile. "And on you, Devon. I don't think I could do this without you."

Their gazes locked and held for several long moments before he looked away. "I'll do everything possible to keep you safe. And I'm glad you have your faith too."

She wished there was something she could say to convince him to put his trust in God, but he clicked on the screen, bringing up another victim's image. Some of these girls were so heartbreakingly young.

The way she once was. If not for Jayme keeping them together after they escaped the Preacher, Caitlyn knew her life would likely have turned out very differently.

And not for the better.

"She's not the one." Her voice sounded weak, so she drew in a deep breath and waited for the next image to bloom on the screen. "No, she's not the victim either."

They searched through photographs for another hour, compiling a list of five women she couldn't completely rule out.

"I need a break." She rose and almost tripped over Snowball on her way to the bathroom. After using the facilities, she washed her hands and splashed cold water on her face.

She stared at her reflection in the mirror, wincing at the dark circles beneath her eyes, accented by her pale face. When she closed her eyes, the images of those missing women flashed in her mind.

Enough. She drew in a deep breath and straightened.

She could do this. She didn't have a choice but to keep going. God have given David the strength he needed to fight Goliath, surely she could manage to look through more pictures.

Easier to do if she didn't keep thinking about what may have happened to them.

It occurred to her that she'd dedicated her life to taking care of animals as a way to escape the horrible things people did to each other. Animals were pure and innocent. They loved unconditionally.

Like Snowball, who'd already captured a piece of her heart.

When she emerged from the bathroom, she saw Devon staring grimly at the computer screen. Despite her inner pep talk, her stomach sank.

"What is it?"

"Nothing." He turned to look at her. "I'm in the process of digging into the backgrounds of the five women who you couldn't say for sure weren't our victim."

"Yeah, so why did you look upset?" She plunked herself down in the seat beside him. "She's dead?"

"Yes, but over a week ago now. So not your victim." He gestured to the screen. "There's sometimes a delay in updating the system."

A delay? She tried not to groan. "In other words, some of these other victims may already be confirmed as dead too."

"Or found," he hastened to add. "But yes. Updating the database isn't the first thing that's done when a woman is located."

Logically, that made sense. Clearly there were other concerns, like understanding what happened and arresting those responsible.

But the news only made her feel more like David holding nothing but a tiny slingshot in her hand. The task before them seemed insurmountable. Especially as there was no guarantee the woman who'd been murdered was listed in the database as missing.

"Let's keep going," she said to Devon.

Dear Lord, give me strength!

IT WAS KILLING him to see Caitlyn so determined to find her victim. As they worked throughout the afternoon, she grew so pale that he feared she'd pass out cold.

But she didn't. Darkness cloaked the motel room by four o'clock in the afternoon, and he decided it was time to take another break.

"Rest your eyes," he advised, closing the laptop. He'd had to connect the power cord as they'd quickly drained the battery.

"We have to find her." Caitlyn's steely determination was admirable.

"She may not be in there," he reminded her gently.

"I know." She sat back and rubbed her eyes. "But if she is, I'll feel terrible if we didn't try to find her."

Devon wished there was a way to shield Caitlyn from all of this. But, of course, her being the only witness made that impossible. Still, he began to understand why Linc had gotten so upset with the repeated attacks against Jayme this past October.

Keeping an emotional distance from Caitlyn was proving impossible. He cared about her, as a friend.

Okay, more than a friend. But that was his problem, not hers.

He rose and moved from the computer, stretching his stiff muscles. He pulled out his phone and called Detective Ernest.

"The ME only has a preliminary TOD," the detective said in lieu of a greeting. "Roughly three to four hours. He won't say anything more definitive until after he's completed the autopsy."

"Three to four hours." Ernest had called him at one thirty to let him know Powers was dead. He and Caitlyn had left the veterinary clinic at ten thirty, maybe ten forty at the latest. "What do you think? Before he shot at us at the clinic? Or after?"

"We don't have any proof the guy who shot at you also killed Powers," Ernest protested. "But if they are the same? I say he did the deed before finding you at the clinic."

He frowned. "The timeline doesn't work. Don't forget he ran us off the road at quarter to ten. I figure he must have been in the area, waiting at the clinic when we arrived."

"I don't know what to tell you." Ernest sounded exhausted. "I knocked on Powers's door at eight thirty. Maybe the killer was already there but hadn't killed him yet. Do you think I like knowing I missed the chance to interview the guy? There's nothing I can do but continue working the case. We have crime scene techs going through Powers's house. Hopefully, we'll find something useful."

"I hope so too." He knew, better than most, that getting evidence wasn't as easy as the TV shows made it out to be. "I'm not trying to be a pest. I just want this guy caught as soon as possible."

"You and me both."

"What did you find outside the veterinary clinic?"

"Nothing yet."

"What about shell casings?" Devon asked with a frown.

"The guy took two or three shots at us. He must have left at least one casing behind."

"Nope. We found the tree where we believe he was stationed, there was trampled snow at the base as if he'd wiped away his boot prints. And I'm thinking he must have picked up his brass too."

That news was hardly reassuring. "Maybe he spent time in the military?"

"Maybe," Ernest agreed. "The thought crossed my mind. Especially as he seems to be good at eluding us." There was a brief pause before he added, "I don't suppose you found anything on the mystery victim."

"Not yet." He glanced over to where Caitlyn was sitting cross-legged on the floor, playing with Snowball. "We're doing our best."

"I know. Keep looking and I'll let you know if we find anything."

"Thanks, Wade." Devon ended the call, frustrated that they weren't any further ahead on the investigation. Ernest was doing his best, but this guy was good.

Too good.

"Time to get back to work?" Caitlyn asked.

"Sure." What else could they do? Sitting around and watching a movie didn't seem right, not when this guy was doing everything possible to get away with murder. "Let's do it."

"Why do you think this guy spent time in the military?"

He shrugged. "It's just a theory. It takes strength to choke the life out of someone, but the way he tried to rear-end us, and then calmly took shots at us after disabling the car? That takes a cool and cunning mind. Plus, he was smart enough to pick up his shell casings and to smudge any hint of footprints in the snow. Shooting Powers, though, takes things to a new level." He couldn't find the words to explain his gut feelings. "I could be wrong. This guy might just be that good."

"But most criminals aren't this good."

"No. Thankfully." He forced a smile. "Come on, let's get back to work." He offered his hand to help her up off the floor. The tingle of awareness wasn't easy to ignore.

He logged into the computer and decided to do a quick check of his email. The most recent unread message was from a strange email address that only had three letters BCF followed by three numbers 826. He was about to send it straight to the spam filter when he caught a glimpse of the first few words.

Give her up.

He clicked and opened the message to read the rest. He stared in disbelief as the email had come from the killer.

Give her up and live. Keep her hidden and die. Your choice.

Devon swallowed hard. How had the guy gotten his email address?

"He messaged you?" Caitlyn's voice came out in a high squeak.

He regretted showing her but reached for his phone to call Ernest. "The killer emailed me."

"What?"

Devon recited the message. "Can one of the tech guys trace this email? I don't know how he got my email address, but this could be the break we need."

"I'll have one of the computer geeks head over to meet you."

"No, I'd rather meet him at the station." Call him paranoid, but he didn't want anyone to know the location of the motel. Not even someone within the Sevierville Police Department. "I'll bring the laptop."

"Okay, that's fine. I'll have someone meet you there in thirty minutes."

Devon disconnected from the phone and glanced at Caitlyn. Her gaze was still riveted on the computer screen.

"He's right, you know," she said softly. "Giving me up would likely save your life."

"First of all, that's never going to happen." He was hurt that she'd even suggest he'd do something like that. "I promised to keep you safe, and I will. Secondly, I don't trust this guy as far as I can throw him. He'd probably kill me anyway. And lastly, he's so anxious to get to you that he's leaving an electronic trail. Don't you see? We may be able to trace this back to him."

"I hope you're right." She shook her head. "I'm not techy enough to hide the source of my email. But many people are." She hesitated, then added, "This guy might know how to do that too."

She wasn't telling him anything he didn't already know. The way this guy had remained a step ahead of them didn't bode well for their ability to track him down.

But they had to keep trying. Giving up wasn't an option.

And neither was handing Caitlyn over as some sort of sacrificial lamb.

"I'm taking this to the police station." He closed the computer and disconnected the power cord from the wall outlet. "You're coming with me."

"Okay. But first we have to give Snowball another dose of her antibiotic." He tried not to groan. "And I'll feed her too. Just in case this takes a while."

"All I have to do is hold her, right?" He bent and gathered the kitten into his arms.

Caitlyn dug the bottle of antibiotics out of her coat pocket. When she had the pill in her hand, she lifted a brow. "Ready?"

He tightened his grip on the squirmy cat. "Go for it."

Thankfully, Caitlyn knew what she was doing. In an instant, she'd forced the cat's jaws open and shot the pill in. The cat seemed to look at him with reproach.

"It's for your own good," he said to the kitten.

Caitlyn shook her head in amusement and opened a can of cat food. It looked and smelled awful, but he knew better than to say anything. "Here you go, Snowball. Enjoy."

Meow. He released the kitten, and she made a beeline for the food.

Caitlyn sighed. "I'm feeling guilty that I haven't had time to look for her owner."

"Once this guy is caught," he assured her. "I'm sure Snowball's owner will understand the delay. Besides, you're taking good care of her."

"True." She reached for her coat. "Let's go."

He kept Caitlyn close to his side as they left the motel. The air had warmed up a bit; it never stayed super cold in this part of the state for long. Once she was seated in the passenger seat, he hurried over to the driver's side.

He took a circular route to the police station. Leaving Caitlyn in the motel alone wasn't an option. At this point, especially with this most recent message from the killer, he wasn't about to let her out of his sight.

Their tech expert was waiting for them. His name was Sam, and he looked younger than Caitlyn.

"Let's see what you have," Sam said, taking the computer from his hands. "Guy has some nerve contacting you."

"No lie." He and Caitlyn sat on opposite sides of the computer guru. "Any idea how he got my email address?"

"Depends, some people use their name, which makes it super easy." Sam turned the computer toward Devon so he could log in, then went to work. "Hmm, yours isn't your name, which should make it harder for him to find."

"This guy knows Caitlyn's name, address, where she works, and my name and address too." Just thinking about it made his blood boil.

"Don't forget, someone searched your house," Caitlyn said. "Maybe he found something there."

"Dude," Sam said in a chiding tone. "If you printed any emails and had them lying around, he easily would have seen your email address."

"I don't usually do that, but maybe." He wanted to smack himself in the head. "Okay, still, we need to know where he is."

"I'm trying." Sam's attention was centered on the screen. "Anyone can make a fake email, but tracing it to an ISP address might give us something to go on."

Please, God, give us something to go on. The prayer slipped into his mind before he could stop it.

"This is interesting," Sam said.

His pulse skipped, and he leaned closer. "What?"

"The ISP on this email indicates it originated in a public place." The kid hit more keys. "It's from a local coffee shop."

Just that quickly the balloon of hope deflated. "So no way to track him."

"Well, we know he was at the coffee shop sixteen minutes ago." Sam looked up at him. "And it's the one located right across the street."

"From here?" He jumped to his feet. "Why would he

choose to email me from a place close to the police station?"

"Hey, I'm just the tech guy," Sam protested. "You're the cop."

Devon was very afraid the killer had lured them to the police station on purpose. Although it didn't make sense. There were more cops coming in and out of the place, especially during shift change.

"I need you to stay here while I run across the street." He stared at Sam. "Sit with Caitlyn, okay?"

"It's no hardship to sit next to a pretty lady," Sam said with a grin.

Ridiculous to feel an ounce of jealousy. Devon hurried outside and crossed the street to the coffee shop. When he went inside, though, there was no one fitting Caitlyn's description of the killer. He went up to the counter and flashed his badge at the barista.

"Did you serve a tall man with dark hair in his early to mid thirties within the past thirty minutes?"

"Yes, why?" She looked surprised by his question.

"What was his name?"

"Devon," she said without hesitation. "I remembered because it's not that common of a name."

The killer had used his name. Devon managed to hang on to his temper with an effort. "Did he use a credit card to pay?"

"No, cash." The barista frowned. "Did he do something bad? Normally, I get a lot of cops in here, so I don't worry so much about getting robbed."

Her statement hit him like a sledgehammer to the head. *She served a lot of cops.*

What if this guy didn't have military background but police background?

This killer could very well be one of their own.

CHAPTER SEVEN

Caitlyn shifted in the hard plastic seat as Sam continued working on Devon's computer. "Where did you learn how to do that?"

"What?" Sam looked at her blankly, then frowned. "Computer stuff? I've been working on computers since I was a kid."

She didn't want to admit that she hadn't used a computer until she'd started high school. And even then, she'd been behind her classmates, thanks to the years they'd lived with the Preacher and out on the streets.

Jayme had insisted Caitlyn go to school every day, even though Jayme herself hadn't had that luxury. Jayme had obtained her GED and had then gone on to become a physical therapy technician. Jayme had also insisted she become a veterinary technician.

Not bad for two foster kids who'd lived on the streets.

"This guy covered his tracks well," Sam muttered. "Smart and cagey."

Great, just what she didn't want to hear. The man

who'd murdered two people was smart and cagey. "But you're better than he is, right?"

"Maybe." Sam continued staring intently at the screen as his fingers flew over the keys. "The ISP originated from the coffee shop, but I should be able to track it back even more to identify the computer itself."

Sounded complicated. "I don't see how identifying the computer will help us find him."

"If I find it while he's online, I might be able to track his location," Sam said. "No guarantee, he could have routed this through a variety of servers."

Since she knew next to nothing about what he was talking about, she fell silent. Footsteps had her glancing over to see Devon striding through the police station, his expression grim.

"Find anything?" she asked.

"The barista remembers a man who fit your description of our perp, but he gave her a fake name." Devon hunkered down next to Sam. "Tell me you have something more."

"Not yet." Sam sounded a bit testy. "These things can't be rushed."

"Sorry." Devon backed off and glanced around the police station. "Caitlyn? Will you come with me for a moment?"

"Sure." She stood and followed as he zigzagged through the room. He went over to stand next to a large bulletin board covered with photos of uniformed officers. "Take a few minutes to look at these cops. Do any of them look familiar to you?"

She frowned, trying to understand what he was getting at. She shifted her gaze from one photograph to the next. "This guy," she said, tapping one picture. "This is Officer Hill, right?"

"How do you know him?" Devon asked.

"He responded to a call back when Jayme was being targeted by the arsonist." She eyed him curiously. "Why are you asking me about these cops?"

"Keep looking," he said, sidestepping her question.

She did as he asked, looking at every single photograph. When she finished, she shrugged. "Some of these cops look familiar, like maybe I've seen them out and about in town, or maybe one of them brought a pet into the clinic. But that's all."

"You don't recognize the man who strangled the woman?" he pressed.

"No." Realization dawned. "Wait a minute, you think the killer is a cop?"

"Either a cop or military, or both," Devon admitted. "I should have suspected someone within law enforcement before now. A cop has access to the DMV database, he could have gotten my license plate number and ran my information to find my address." He stared at the sea of faces on the wall. "I was really hoping this guy worked here in Sevierville, but if you don't see him, he might be with another police force nearby."

A cop. The memory of the man's cold hard face as he strangled the woman to death flashed again in her mind. The thought that the same man who'd brutally murdered a woman was also a guy who'd taken an oath to serve and protect made her feel sick to her stomach.

"Is there a way to get photos like these"—she waved a hand at the bulletin board—"from other police departments nearby? Knoxville? Or even Pigeon Forge?"

"I hope so." He drew her back to Sam. "Stay here, I'm going to make a few calls."

"Wait, I think I have something," Sam said. The guy's

eyes gleamed with anticipation. "He's online right now, and I'm close to pinpointing his location."

"Where?" Devon leaned down to look over his shoulder. "I really need to know where he is."

"Hang on," Sam muttered. "Okay, here. Looks like he's using the Wi-Fi at a place called Cooper's Town."

Devon used his phone to plug the information in. "That's a restaurant in Knoxville."

Caitlyn swallowed hard. "You can't call the Knoxville police to get him, not if he might be one of them."

"A cop?" Sam gaped in surprise but then nodded. "Makes sense."

Devon lifted his phone to his ear. "Detective? We have reason to believe our guy is at a restaurant called Cooper's Town in Knoxville. I want you to meet me and Caitlyn there."

She shouldn't have been surprised Devon intended to take her along since she was the only one who could identify him. She knew it was important to get this guy behind bars where he belonged. Still, her stomach did a little flip at the possibility of seeing him again.

Doubts assailed her. What if her memory wasn't as clear as she'd portrayed? What if she didn't recognize him?

"Good. We'll see you soon." Devon lowered his phone and took her arm. "Let's go."

They only took two steps when Devon suddenly stopped and turned to face Sam. "Hey, what do you drive?"

"Me?" Sam looked surprised by the question. "I have a Chevy Tahoe that's like ten years old. Why?"

"I want you to switch with me." Devon went over and set his keys on the desk. "Take the dark gray SUV with the muddy license plate if you need to go."

"Okay." Sam readily dug his keys from his pocket. "It's rusty but runs pretty good despite the high mileage."

"Did you park out back?" Devon asked. When Sam nodded, he turned to take her arm. "Time to hit the road."

She wasn't sure why Devon had wanted to swap vehicles, it seemed overkill, but she didn't argue. She accompanied him outside and easily found Sam's Tahoe. It was a messy disaster inside, but she gamely pushed the fast-food wrappers off the seat and onto the floor so she could get in.

"Bring up the address for Cooper's Town on your phone and use the map function to give me the route," Devon directed. "This car doesn't have a built-in GPS."

"Okay, the restaurant is sixteen minutes away." She began rattling off directions from her phone. "Turn right at the next intersection."

The roads were fairly deserted this time of the evening on a Sunday. Devon glanced at her. "Once we get to the restaurant, I want you to stay behind me at all times."

She gave a jerky nod. "I'll try, but I'm sure you'll want me to look at all the restaurant customers to see if I can identify him."

"Yes." Devon's expression was grim. "But I also don't want you to be in danger, Caitlyn. So the minute you spot him, I want you to get someplace safe."

"It's strange to think of him sitting there, eating dinner," she said as a way to distract herself from the gravity of their mission. "I mean, he's been coming after us nonstop."

"I know." Devon pushed the Tahoe's accelerator. "I wish I'd thought of the possibility of his being a cop sooner. He had the audacity to give the barista my name for his drink order."

She sucked in a harsh breath. "He did?"

"Yeah. As if he knew I'd track him there." Devon shook

his head. "I'm concerned that as a cop he might be able to cover up any report of a missing woman. Especially if she happens to be his girlfriend."

Horrible thought. "It also may be too early for anyone else to know she's gone," Caitlyn said. "But maybe tomorrow things will change. If Jayme was gone and didn't answer my phone calls or text messages, I wouldn't just stop at the police station. I'd use every media source I could find to blast her photo everywhere."

"Good point," Devon agreed. "Although if he has her phone, he could respond to text messages, pretending to be her."

"True." Caitlyn dragged in a deep breath. She couldn't fail in her task of identifying this guy.

Please, Lord, grant me the strength and wisdom to find him!

Devon arrived at the restaurant a few minutes early. The place wasn't anything fancy and didn't offer live music the way so many of the pubs and taverns did in Tennessee. She watched Devon scour the parking lot. "I don't see Detective Ernest."

"I'd like to go inside right away to get started," she said, trying to sound confident despite her trembling fingers.

Devon hesitated a moment, then nodded. "Okay, but remember the plan to stay behind me."

"I'll try." She'd do her best, but Devon was hardly invisible. She'd need to have an unobstructed view of the restaurant customers.

Devon led the way inside. There was a small entryway with a hostess stand where guests were expected to wait to be seated. Devon tried to push her behind him, but she stayed near his side as she glanced from one group of occupants to the next.

She took her time, scared she'd make a mistake and identify the wrong man. But after a few minutes, she felt certain he wasn't there.

"See him?" Devon asked in a low voice.

"Not yet." There was a cluster of tables off around the corner that was outside of her vantage point.

"Dinner for two?" the hostess said as she came forward with two menus.

"Ah, we're just looking for someone." Devon moved over to the side, which only made it more difficult to see the corner of the room.

"Oh, who are you joining?" the hostess asked.

"If you don't mind, we're just going to check and see if our friend is here." Caitlyn smiled at the woman and tugged Devon's arm. She tipped her head to the side. "He might be in the corner back there."

Devon swept his gaze over the room, then went around the hostess stand to a spot where the corner tables were visible. There was a foursome sitting back there, two men and two women.

Caitlyn's heart sank. "I was wrong. He's not here."

Devon grimaced and nodded. "I was afraid of that. Let's head back outside to meet with Detective Ernest."

Caitlyn nodded, knowing it wasn't her fault the murderer wasn't there the way Sam had told them.

Still, it was a setback in the investigation. She found herself wondering if the murderer had been there but already left.

Or if he'd never actually been there at all.

DEVON QUICKLY ESCORTED Caitlyn to Sam's Tahoe. He called the computer guru, who answered on the first ring. "He's not here, Sam. Are you sure you didn't give us the wrong place?"

"I'm telling you, he was using the Cooper's Town Wi-Fi," Sam insisted. Then he added, "I guess he could have accessed their free internet from outside the building rather than being inside."

Devon gripped the steering wheel tightly, realizing he should have thought of that. He disconnected from the call without saying anything more and began looking at the vehicles around them.

"Write down as many license plates as you can," he told Caitlyn. He twisted the key in the ignition and shifted the truck into drive.

With her phone in hand, Caitlyn began tapping in license plate numbers. Maybe this was a futile effort, the killer may have seen them drive up and took off while they'd been inside.

Then again, if that was the case, why hadn't the perp made another attempt to kill them?

Maybe his last-minute idea of borrowing Sam's Tahoe had caused some confusion. The killer wouldn't have expected them to show up in a rusty old truck.

Score one for the good guys.

He tried not to let the loss of nailing this guy get to him. Finding him at the restaurant hadn't been a sure thing. The twelve minutes it had taken for them to get there was ample time for the guy to have escaped.

"So close," he muttered, taking the Tahoe around the parking lot to get all the license plates.

"I know." Caitlyn's tone held regret. "I wish we'd have

found him, too, but don't worry. I'm sure we will get him very soon."

He was ashamed of his attitude. Caitlyn had the most to lose, and here she was consoling him. His phone rang, and he saw Ernest's truck pull into the parking lot. "He's not here," Devon told him. "We're taking down license plates, just in case. But he's not inside as a customer."

"What if he's a staff member?" Ernest asked.

"We can check, but my theory is that this guy has a background that includes either military or law enforcement training. Not a bartender or dishwasher for a restaurant."

"A cop?" Ernest whistled. "Makes sense, especially the way he cleaned up both crime scenes, the one in the woods and the one outside the veterinary clinic. But I feel like we should check the employees anyway."

"Okay." Devon understood the detective's need to be thorough. In fact, he should have considered that too.

Maybe he wasn't detective material. It seemed he'd done nothing but underestimate this guy from the very beginning.

"We're going back inside?" Caitlyn asked, looking up from her phone.

"Yes." He parked the Tahoe and shut down the engine. "We're going to check the kitchen staff and the other employees."

"I don't think he works here," she said. "But I guess it can't hurt to check."

"Agreed." Then he frowned. "I'm going to wait here while you go inside with Detective Ernest. I think it's important to make sure there isn't another attempt to sabotage our vehicle."

She hesitated, a flash of fear darkening her blue eyes,

before she nodded. "Okay. That makes sense."

It was the smart, logical thing to do. So why was he so loathe to let her go off without him? They were going into a restaurant, not a pit of viper snakes. And while Ernest wanted to be thorough, he agreed with Caitlyn.

No way did a killer this smart work in a dive restaurant.

But he'd been there. Either inside or lurking around outside the place.

"Rainer." Ernest startled him by rapping on the window. "You ready?"

He lowered the window. "Take Caitlyn inside with you. I'm going to keep an eye on things out here."

Ernest turned to glance around, then shrugged. "Okay."

Caitlyn slid out of the passenger seat and went around the Tahoe to join Ernest. It took all of his willpower to let her go inside the restaurant without him.

He scrubbed his hands over his face. This was not good. He was more than a little emotionally involved with Caitlyn Weston.

Ernest and Caitlyn would be in and out within ten minutes. Maybe less. He forced himself to sweep his gaze over the area, looking for any sign of their killer.

A flash of light through the trees snagged his attention. Headlights from a car on the road? His heart pounded in his chest as he tried to imagine the layout of the area. He wasn't very familiar with Knoxville, but it couldn't be that much different from Sevierville, where he lived. The roads often twisted and turned through the mountains.

He pulled out his phone and used the map app to bring up the restaurant. From there, he zoomed out to get an idea of the nearby highways and roads.

There was no road in the area where he'd seen the light. Only woods.

Was it possible there was a house tucked back there? Maybe. A long driveway wouldn't show up on a map.

The light winked again, taunting him.

He told himself to get a grip. That a light in the trees didn't mean the killer was out there, trying to draw him out.

Then again, what if it was?

Reaching up, he removed the dome light cover and bulb. If he was going out to investigate, he didn't want to broadcast his intentions. He slid out from behind the wheel, softly closing the door. He pulled his weapon and lightly ran into the woods, using the evergreen trees and large tree trunks for cover.

Deep down, he almost hoped the killer was out there, waiting for him. He wanted this guy arrested and tossed in jail as soon as possible.

And if it wasn't? Some might say he'd look like a fool for traipsing through the woods searching for some distant light, but those same people hadn't had their house searched or been shot at outside a veterinary clinic.

In his humble opinion, the killer's taunting him by sending an email and using his name at the coffee shop meant he was growing desperate.

As if the guy knew that he'd have to go through Devon in order to get to Caitlyn.

He was about twenty yards into the woods when the light flashed again. Definitely closer this time. Devon dropped down behind a large tree, wishing he had night vision goggles to see who or what was up ahead.

The clock ticking in his head told him he didn't have much time before Ernest and Caitlyn would be finished inside the restaurant. He darted out from behind the tree to another, and yet another, each time making his way deeper into the woods.

He felt certain he was getting close to the source of the light. Unless the killer had seen him and was also moving away.

The possibility was sobering.

Devon ran for another ten yards before dropping behind another evergreen tree. The air was growing cold enough that he could see his breath.

"Devon!" Caitlyn's shout had him snapping his head around to the parking lot. "Where are you?"

Just then, he heard the snap of a tree branch. Swinging back around, he saw the light bobbing in the distance.

The killer was getting away!

Ignoring Caitlyn, he jumped up and ran after the bobbing light. It quickly vanished, as if the killer suspected Devon was close by.

He couldn't let this guy get away. Rather than using the trees for cover, he simply ran through the woods as fast as he could, which wasn't quick enough given the fallen logs and thick brush obstructing his path. He couldn't see anyone up ahead, but that fact didn't cause him to slow down.

If anything, he pushed himself harder. And once again found himself praying for strength and guidance.

Show me the way, Lord!

He tripped over a fallen log but managed to stay on his feet. As he pushed deeper into the forest, he grew convinced the killer was out there.

Crack!

The echo of gunfire was so loud he tripped again, this time hitting the ground hard. Instantly, he was up on his feet, darting to the closest tree to use for cover.

"Devon!" Caitlyn's cry held a note of panic. He wanted to call out to her, but he couldn't give his position away to the gunman.

Get her out of here, he silently urged Ernest. *Keep her safe!*

For long moments there was nothing but silence. Devon knew he couldn't hide behind the tree forever, so he edged around it and dashed toward another tree.

But there were no more flashes of light. And oddly enough, no more gunfire.

Where had he gone? Was he lying in wait for Devon to come closer?

Pulling his phone from his pocket, he used his jacket to cover the light the best he could as he called Ernest. "Is Caitlyn safe?"

"Yes, I have her in my vehicle," Ernest said calmly. "Did you find the shooter?"

"No." His shoulders sagged in relief at knowing Caitlyn wasn't hurt.

"I'll call the Knoxville PD for backup," Ernest said.

"Don't do that, there's a chance this guy is part of the Knoxville PD." Devon drew in a deep breath. "Just stay where you are, I'll join you shortly."

"Got it," Ernest agreed.

Devon carefully turned and went back through the woods the way he'd come. Soon he was back at the Cooper's Town parking lot. He rushed over to Ernest's car and climbed in. He grimly met the detective's gaze. "I lost him."

"At least you're not hurt," Ernest said. "You shouldn't have gone after this guy through the woods without backup."

Devon ignored the rebuke. Ernest was right to a certain extent, but these were hardly usual conditions.

And this perp wasn't your average killer. He forced back the bitter taste of failure at knowing he'd lost him again.

CHAPTER EIGHT

Caitlyn kept her thoughts to herself as Devon drove them back to the police station. She hated knowing that he'd gone into the woods after the killer and had been shot at. Yet she couldn't deny sincerely wanting this guy found and arrested.

"What if he'd killed you?" Okay, maybe she couldn't keep her thoughts to herself. She twisted in her seat to face him. "You promised to protect me, but you can't do that if you're injured or worse."

"I'm fine," he said, barely looking at her. "And I'm not going to apologize for trying to get him."

She threw up her hands. "Yeah, okay. So once he takes you out of the picture, he'll find it easier to come after me." It was disconcerting to realize she cared more about Devon than herself. "I have a better idea. Use me to draw him out of hiding. That way this nightmare will be over once and for all."

"Don't be ridiculous." There was a hard edge to Devon's tone. "No one is going to use you as bait."

"Isn't that what you just did? Used yourself as bait?"

She turned back to face the windshield and crossed her arms over her chest. Knowing how close he'd come to being shot had shaken her.

Badly.

"I'm a cop. Going after bad guys is what we do."

"Not without backup." She'd heard the detective's comments and had wholeheartedly agreed with them. After a few moments of silence, she added, "I care about you."

He reached over and rested his hand on her knee. "I care about you too."

It wasn't an apology, but she nodded and covered his hand with hers, needing the physical connection.

She understood Devon's job was dangerous, that he put his life on the line every time he walked out the door in his uniform.

The impact of his chosen career hadn't really hit her until now.

"Are you hungry?" Devon's question pulled her from her troubling thoughts.

"I could eat." She wasn't that hungry, but old habits die hard. When she and Jayme were struggling to make ends meet, they hadn't always had enough to eat. Living in the lean times had taught her to never turn down a meal. Like, ever.

"We'll grab some takeout, if that's okay." He glanced at her. "You can choose this time."

"Chinese?" She arched a brow.

He chuckled and nodded. "Sure. There's a great Chinese place not far from our motel."

Life was too short to stay angry, so she decided it was time to put their future in God's hands. Always easier said than done, but she lifted her gaze to the starry sky and did her best to honor Him.

Please continue to watch over us, Lord. Let Thy will be done. Amen.

When they arrived at the police station, Detective Ernest pulled in right behind them. The three of them walked inside together.

Sam was still seated at the same desk, working on Devon's computer. His expression was a mixture of frustration and chagrin when they approached. "I'm sorry, I thought for sure he was inside the restaurant."

"He was there," Devon said with a sigh. "I followed him into the woods. He led me deeper into the brush, then took a shot at me. I didn't get a look at the perp to know for sure that he's our guy, but I doubt anyone else would randomly fire a weapon at a stranger."

Sam grimaced, then pointed at the computer. "I attached some tracking software to your email. If he reaches out again, we'll be able to find him."

Caitlyn didn't bother to point out that tracking the guy wasn't as easy as the tech guru made it sound. The near miss in the woods was proof of that.

"Here're your keys." Devon tossed them at Sam who missed catching them.

"Hey, I didn't even get a chance to drive your car," he protested.

"Not my fault," Devon teased as Sam handed over the keys. Then his expression turned serious. "Be careful, Sam. I'm pretty sure the shooter saw the Tahoe."

"If that's true, then Sam needs an unmarked car too," she said with a frown. The last thing she wanted was for someone else to be hurt by this guy.

"Uh, is that allowed?" Sam looked at the detective.

"Sure, help yourself," Ernest said. "I think Devon is

right. If he saw your vehicle, it's not safe to drive it. We cannot keep underestimating this guy."

Devon helped Sam sign out a car, then packed up his computer and turned to her. "Ready to go?"

She nodded. Devon led her outside to the dark gray SUV. As before, he used a circular route back to the motel, taking several turns in order to make sure they weren't followed. When she saw the Chinese restaurant, she gestured to it. "Are you planning to stop?"

"No, I think I'll have them deliver." He shrugged. "I'll feel better once we're locked inside the motel room."

"Me too." Her stomach growled as she imagined the tangy soy sauce and egg rolls. "I'll eat just about anything, but sweet-and-sour chicken is my favorite."

The corner of his mouth tugged up in a smile. "Duly noted."

He pulled into the motel parking lot but then drove all the way around to the back of the building. She swallowed hard, realizing he was doing his best to prevent the killer from finding them.

From finding her.

The email message had rattled her. Devon wasn't going to turn her over to some madman, and anyone who knew him would understand that. But he was just as much of a target now too.

Maybe more so. Because as she'd said earlier, if the killer eliminated Devon, he could easily get to her.

They headed inside the motel. While Devon placed their food order, she spent some time with Snowball. The cat's laceration looked good, thanks to the antibiotics along with the thick red collar that prevented her from licking it.

"I'll post something on social media tonight about your owner, okay?" She stroked the cat's soft fur.

"Not tonight or any night until this guy is caught," Devon said from the connecting doorway between their rooms. "Promise me, Caitlyn."

"Okay, you're right." She sighed, feeling that Snowball's owner was probably worried sick about the feline. Then she frowned. "Couldn't I just look at the social media sites to see if anyone has posted a missing cat? I won't reach out to the owner until the danger is over, but it would be nice to know who she belongs to."

He nodded slowly. "Okay. Although I was thinking we should keep going through the list of missing women to find our victim."

"You're right, that's more important." Still, she pulled out her phone and quickly pulled up a social media site. After typing "Sevierville - missing white cat with blue eyes" in the search bar, she scrolled through to see if there were any hits.

Surprisingly, there weren't.

She wondered if the cat had wandered far from home. That wasn't normal, mostly cats tended to stick around the area from where they lived.

Devon booted up his computer and logged in. "Caitlyn? Are you ready?"

"Yes." She set her phone aside, gave Snowball one last rub, then joined him at the small table. Looking at these women's photos was depressing, but she told herself to get over it.

The dead woman's family deserved to know what had happened to her. It was up to Caitlyn to find and identify her.

She and Devon worked until the Chinese food arrived. He paid for the order while she cleared off the table.

Together they unpacked the white containers until everything was ready.

Taking her seat, she bowed her head to pray, but Devon surprised her. "I, uh, would like to pray too."

Her heart filled with joy at his words. Smiling, she took his hand. "Dear Lord, we thank You for keeping us safe in Your care today. We ask that You continue to watch over us and guide us as we search for this dangerous criminal. We also thank You for this food we are about to eat. Amen."

"Amen," Devon responded. He didn't release her hand until she looked up at him. "I prayed earlier, while I was tracking the killer through the woods."

"Oh, Devon. I'm thrilled to hear that." She hoped she'd contributed to his opening his heart and mind to God. "I know God is watching over us in this difficult time."

"I have to admit, it helps to know that." Devon lifted her hand and lightly kissed her knuckles before releasing her. She felt the brush of his lips on her skin all the way down to her toes. "And maybe once this is over I can learn more."

"I'd love to teach you." She wanted to jump up and dance around the motel room with happiness, but she managed to refrain. Instead, she picked up a plastic fork and dug into the container of sweet-and-sour chicken.

"I need to call my boss when we're finished here," Devon said, changing the subject. "I have to let him know that I won't be in to work in the morning."

"Are you sure you won't get in trouble? I thought maybe you'd have some other cop friend of yours watch over me." She hadn't been looking forward to being in the motel without Devon, but his job was important.

"I was thinking of asking for a colleague to help out," he admitted. "Getting that email from the killer changed my mind."

The sweet-and-sour chicken lodged in her stomach. Of course, Devon wouldn't want to put anyone else in danger.

Bad enough she'd dragged him into the line of fire.

"I wish I'd gotten a closer look at him," Devon said. Then he frowned. "We could have you work with a sketch artist."

"I'm willing to do whatever you need," she agreed. She thought of her foster brother Cooper who was an amazing artist. "Maybe the bartender at Flannery's could work with one too. Between the two of us, we should be able to create a likeness."

Devon nodded thoughtfully. "I'll put a request in to my boss when I talk to him."

It felt good to be able to contribute something constructive to the investigation. She only hoped she could do a good job. She'd recognized him as being at Flannery's, but she hadn't memorized his facial features.

To be honest, she could tell animals apart more easily than people.

Yet she'd find a way to do this. Failing Devon wasn't an option.

DEVON FINISHED HIS MEAL, feeling oddly content. Having Caitlyn someplace safe was a huge relief. But he also was coming to realize the odd sense of peace was coming from God.

Strange because he'd never really felt God's presence. Then again, he'd never been in a killer's crosshairs either.

As a cop, he was often walking into potentially dangerous situations. Yet Sevierville wasn't a hotbed of crime. They had their usual drunk and disorderly conduct, their drugs, some physical abuse along with some burglaries. But straight up murder was a rare occurrence in their small tourist town.

First an arsonist had come after Jayme, now this. Granted, Caitlyn's witnessing a murder wasn't something she'd done on purpose. If not for the cat, she wouldn't have gone into the woods at all.

And if she hadn't witnessed the murder, he wasn't sure the police department would ever have known about it.

He kept coming back to how the killer had moved the body. Where had he taken his victim? To another dense, wooded area?

If that was the case, they likely wouldn't find her until spring, when some poor hiker stumbled across her bones.

The grim thought gave him a renewed sense of purpose. He had to find this killer, and soon.

"Is there a way to find out if more missing women have been added to your database?"

"I did that earlier, nothing has been added today yet."

"Oh." She sighed. "I know it's possible this guy may have taken this woman a long time ago, but looking at women who have gone missing six months ago seems like a stretch."

"I know." He finished his food and pushed his empty paper plate aside. "Unfortunately, I don't have a better plan. Other than setting you up with a sketch artist first thing in the morning."

"Okay." She finished her meal too. "We can keep going through the photos if you think it will help."

He glanced at the computer she'd set on the bed. "No, I don't think that's necessary. You're probably right that our victim didn't go missing six months ago. If we had a name, even a first name, we could search social media sites. But without that . . ." He shrugged.

"Maybe the crime scene techs will come up with something."

"I hope so." The partial boot print was all they had at the moment, and that wasn't nearly enough. It would help once they had a suspect in custody, but finding some guy who wore a size twelve shoe would be more impossible than finding a diamond in the rough.

Diamond. What on earth made him think of that? He glanced at Caitlyn's ringless hands.

He gave himself a mental shake. Lack of sleep was making him loopy.

"Yeah, me too." Caitlyn rose and began clearing the empty containers away. He consolidated the leftovers into three boxes. "We don't have a fridge, but I'm sure you'd rather not throw this away."

"No need to waste food," she hastily agreed. "I'll use the ice bucket to help keep them cold."

When she'd finished, he tucked his computer under his arm. "No more working tonight, we need to get some rest. Tomorrow will likely be a long day."

"Yeah." She lifted her phone. "I'm going to call my boss, let him know I can't come in."

"I'll call my boss too." He paused, then added, "Good night, Caitlyn."

"Good night."

He stepped across the threshold of their connecting doorway. He wasn't looking forward to calling his captain, but he couldn't put it off much longer.

"Barstow," his boss answered curtly.

"It's Rainer. I have a situation you should probably know about." He went on to explain about Caitlyn's witnessing a murder and the events that had taken place thus far. Barstow remained silent until he was finished with the entire story, including the way he'd followed the killer into the woods outside the Cooper's Town restaurant in Knoxville.

"That's more than a little situation," Barstow said with a sigh. "That's a whopping mess."

"Yes, sir." Devon was banking on the fact that Barstow knew Linc well enough that he wouldn't risk harm coming to Jayme's younger sister. "I won't be able to make my shift tomorrow. I can't leave Caitlyn unprotected."

"Yeah, I get that." Barstow fell silent for a moment. "Have you reached out to Linc?"

Devon winced. "Not yet. Caitlyn doesn't want to ruin their honeymoon."

More seconds ticked by as his boss considered the pros and cons of that decision. "We can discuss it more tomorrow," Captain Barstow finally said. "Nothing to be gained by calling them now."

"That was my assessment as well. Oh, I do think it would help to have Caitlyn work with a sketch artist. The bartender at Flannery's too. They're the only ones who can identify the killer."

"Sounds good. I'll make some calls, see if we can get Trainer to come in."

Trainer was in his midsixties and retired from his job. He did sketch work for the Sevierville PD on a part-time basis. "That would be great, thanks."

"Rainer?"

"Yes, sir?"

"Be careful. It sounds like this guy is coming after you now too."

"He is, but he wants Caitlyn more. My priority is to

keep her safe."

"Good plan. And don't forget to call the dispatcher about you being off in the morning," Captain Barstow added before disconnecting from the call.

His boss had taken the news better than he'd expected. Devon knew that was mostly because Caitlyn was Jayme's sister. The PD worked closely with Lincoln Quade on arson cases. Clearly, his boss didn't want to take a chance with Caitlyn's life.

Devon made the call to the dispatcher, then washed up in the bathroom. He was physically and emotionally exhausted but sensed sleep wouldn't come easily.

Still, he did his best. Stretching out on the bed, with the blanket thrown over him, he closed his eyes and tried to still his racing mind.

He found himself listening intently for sounds indicating Caitlyn was awake too, but the connecting room was quiet.

Should he go ahead and call Linc on his own? He'd rather the call came from him than from his boss. Yet he'd also prefer to have already arrested this guy before bothering Linc and Jayme on their honeymoon. That way, they could stay and enjoy their week away.

He must have dozed because a muffled cry woke him. He shot out of bed and rushed through the connecting doorway to find Caitlyn in the throes of a nightmare.

"No, stop, please . . ." she whimpered.

"Caitlyn." He reached out to lightly touch her shoulder. "Wake up, you're having a bad dream."

"What?" Her eyes opened, and she stared blankly at him for a long moment before registering what had happened. She sat up and pushed her hair from her face. "I'm sorry I woke you."

"Don't be, I'm just glad you're okay." He'd lost ten years off his life thinking the killer had gotten to her. "Do you have these nightmares often?"

"No." She sighed and gathered Snowball close. "Must be the stress of the day catching up with me."

"I'm sure." He was glad she didn't suffer nightmares on a regular basis. "Nightmares about the Preacher?"

She shrugged. "Yes, normally, only this time the dream didn't make any sense. The Preacher was younger and looked more like the guy I saw in the woods."

"I'm sorry to hear that." He hated feeling so helpless. If he could erase all the bad memories for her, he would.

"It's fine." She offered a wan smile. "I'm nervous I won't be able to provide enough details for the sketch artist."

"You're going to be great." He sat beside her and lightly draped his arm around her shoulder. "God will guide you."

This time her smile lit up her face. "You're right, He will. Thanks for the reminder."

The urge to kiss her was strong, but he told himself this was not the time to take advantage of the situation. He eased away and stood. "If you need anything, let me know."

"Thanks, Devon." She stroked the cat. "I'm sure I'll be fine."

"Okay." He glanced at her one last time before heading into his own room. Her beauty and innocence shined like a beacon, and it was incredibly difficult to keep his distance.

Linc's younger sister, remember?

Despite his lingering desire for Caitlyn, he managed to fall back to sleep. He wasn't sure how much time had gone by when he heard his name.

"Devon."

Caitlyn's voice had him vaulting off the bed, glancing around wildly for the threat. "What?"

"I found something." She held up her phone.

He willed his heartbeat to return to normal. Swallowing hard, he gestured for her to take a seat. "Okay, what did you find?"

"This." She handed him her phone.

He stared at the smiling woman holding a white cat. "You found Snowball's owner?"

"She's the missing woman." Caitlyn's eyes were wide as she looked at him.

"You mean, this is our murder victim? Stephanie Phillips?" He was shocked. "Are you sure?"

"Positive." Caitlyn gestured to the picture. "I couldn't sleep after my nightmare, so I was searching for white cats that were missing. Stumbling across this photograph was a shock. It never occurred to me that she and Snowball might have been together that night. Especially since the cat ran across the road from the opposite direction from where the woman and the man were struggling."

He blinked and stared down at the photograph again. There was no denying the cat looked like Snowball. And the smiling woman with long dark hair also fit Caitlyn's description of the victim.

"This is huge, Caitlyn. Now we can look at her known associates, see if we can pinpoint the man who killed her."

"I already looked through her social media pages but haven't seen her with the man in the woods."

The news was disheartening, but he refused to feel let that fact get him down.

Stephanie Phillips had to have met the killer at some point. He just had to figure out when and where. They were one step closer to getting this guy.

He could feel it.

CHAPTER NINE

___

Caitlyn was both happy and sad that she'd found Snowball's owner. The cat probably didn't belong to anyone now that Stephanie was dead, although she didn't know that for sure. If Stephanie didn't have family, she wanted to adopt the cat. Yet seeing the young woman's smiling face on social media made her chest hurt.

Stephanie hadn't deserved to be murdered. Especially by the guy who must have been some sort of boyfriend. Those open blank eyes would haunt her for a long time.

Odd that she hadn't seen the guy, or any significant others, on social media. Not that everyone used it, she hadn't until recently. When she and Jayme had gotten their first phones, they were cheap disposable ones. It was only when Jayme graduated and started working full time as a physical therapist assistant that they'd gotten better phones. And even then, living with the Preacher had made her wary about using social media. She didn't remember her real parents, only a series of foster homes, the last one with the Preacher being the worst.

She owed Jayme so much. Which is why she hadn't wanted to bother her sister on her honeymoon.

Maybe now that they had the victim's name, she wouldn't have to. Hopefully, this clue would help Detective Ernest find this guy. She couldn't bear to think of him getting away with what he'd done.

"Try not to worry, we're one step closer to finding the killer." Devon's voice interrupted her thoughts. The way he was so in tune to her emotions only made her like him more. So different from Nate, who'd been completely clueless. "Thanks to you, we have a name. We'll get this information to Detective Ernest first thing in the morning."

"I was so shocked to find it," she admitted. "Maybe I should have considered the woman was Snowball's owner right away."

"There's no way you could have known," Devon assured her.

"I wonder if that guy hurt the cat." Caitlyn scowled. "If he did, can you add that to his list of crimes?"

"We can, but you know it won't hold much weight. Not like two murders and the attempted murder of us."

"I guess." Logically, she knew he was right. Stephanie and Nate were the true victims. Snowball had been smart enough to get away. And maybe the feline had doubled back to find her owner. Caitlyn looked back down at the picture on her phone. "The detective will contact her family, right?"

"Absolutely." Devon raked his hand through his hair, then stood. "I'll text him, see if he's awake. If he doesn't respond, though, we'll have to wait until the morning to follow up on this."

"I understand." Truthfully, it was two thirty in the morning, so of course waiting until daylight was the right thing to do. Cops didn't routinely go knocking on people's doors in the middle of the night for an interview.

Devon sent the text, then sat at the small table as he waited for a response. "Looks like you'll be able to keep Snowball."

"Maybe. Unless Stephanie's family wants her." The thought was depressing, but she couldn't complain as long as Snowball went to a good home. It wasn't as if there wouldn't be other strays for her to take in.

Her landlord didn't allow pets anyway. Although what the guy didn't know wouldn't hurt him.

"I didn't see other family members listed on her social media, did you?" Devon pulled up his phone and typed in Stephanie's name. "I think it's strange no one reported her missing."

"Maybe by tomorrow, when she doesn't report in for work." She sat next to him, watching as he scrolled through Stephanie's social media. "I noticed she works at the local health clinic."

"I see that." He glanced up at her. "That's another place Ernest can go to get information on Stephanie. Someone she works with likely knows who the killer is."

She nodded, thinking that there was a really good chance that Devon and Detective Ernest would have this guy behind bars by lunchtime. "Maybe I can still go to work tomorrow."

"Not until we've arrested him." Devon frowned at his phone. "Still no response from Ernest. Let's try to get some sleep."

"Okay." Sleep wouldn't come easy, first because of the nightmare and then again because she'd stumbled across the victim.

God was doing His part in showing them the way.

Caitlyn knew that she and Devon needed to continue to trust in His guidance.

"Caitlyn?" Devon's low husky voice sent shivers of awareness down her spine.

"Oh, sorry." She flushed, realizing she was sitting in his room. She hastily rose to her feet, the same time he did. They were mere inches from each other. She told herself to turn and leave, but instead she stepped closer and hugged him.

He didn't hesitate to pull her close. "Hey, are you okay?"

She nodded, inhaling his unique musky scent. "It's all catching up to me," she said, her voice muffled against his T-shirt.

"You've been awesome through this, Caitlyn." He lightly stroked her hair. "Most women would have fallen apart long before now."

She had fallen apart, but Devon had been there to prop her up. "Thanks, but I know I'm not as strong as Jayme."

"You are," Devon countered. "Look at where you are now, despite the hardships you've faced. You and Jayme are both the strongest women I've ever known."

His sweet words soothed her soul. She clung to him a moment longer before forcing herself to step away. "Thanks, Devon."

"Caitlyn." He brushed a strand of hair from her cheek. "You don't have to thank me for holding you. It's my pleasure."

She stared into his light gray eyes, wishing for something she couldn't have. Their brief kiss aside, Devon mostly treated her like a younger sister. Probably because of his relationship with Linc.

He might have noticed her longing for more because he slowly lowered his mouth to hers and kissed her.

Like before, his kiss swept her away. She melted against him, pouring her feelings into their embrace. He was passionate and sweet, kissing her with desire without being demanding.

When he finally lifted his head, his chest rising and falling with rapid breaths, he continued to hold her, resting his cheek on her hair. "I shouldn't have done that," he whispered.

"Why not?" Irked by his comment, she pulled away to look up at him. "I'm not a young kid, Devon."

He winced. "I know, but you're a lot younger than me."

"Am I?" She lifted a brow. "Maybe in actual years, but not in life experience. I'm not some naïve woman who doesn't know what she wants."

"Caitlyn." He took a hasty step backward, nearly tripping over the chair. "You haven't been in a serious relationship."

"How do you know?" The fact that he was right stung. "I dated Geoff during my senior year of high school. Maybe that doesn't count for much in your opinion, but it's enough for me to know that I prefer animals to men. Especially now."

"I'm sorry. I shouldn't have said that." Devon was obviously trying to backpedal. "I care about you, Caitlyn. I don't want to hurt you."

Too late, she thought wearily. Then she tipped her chin. "And maybe I don't want to hurt you either. Good night."

She turned and walked through the connecting doorway to her own room. Devon didn't try to stop her.

Her lips still tingled from his kiss. Yet she knew that no

matter what she wanted, once the killer was arrested, she wouldn't see Devon again.

———

HE KICKED himself for being a jerk. Twice now, he'd kissed her, then pushed her away.

Normally he didn't have this much trouble controlling himself. He could blame it on the danger surrounding them or the forced togetherness, but deep down, he knew the truth.

He was falling in love with Caitlyn.

Linc would not be happy. And honestly, there was a part of him that worried Caitlyn wasn't ready for a relationship. She was only twenty-three, she had her whole life ahead of her. He'd already failed at one relationship with Sabrina. Heading down that path again was not at the top of his list of things to do.

Her comment about not hurting him hit home. Because he knew she was right. He already cared for her, but as soon as she realized there was no reason for her to settle for a guy like him, she'd likely move on.

Whatever. He needed to keep his head screwed on straight. No more holding and kissing Caitlyn.

Since Ernest hadn't texted him back, he tried to get some sleep. No easy feat with memories of Caitlyn's embrace fresh in his mind.

He tossed and turned, dozing in fits before he finally crawled out of bed at six. He showered, changed, and made coffee. There were no sounds from Caitlyn's room when he listened at the connecting door, and he hoped she was getting more sleep than he had.

This would be the day they'd find and arrest the killer.

Learning the victim's identity would likely give them additional clues to go on. If the perp had been a current or former boyfriend.

If not? He grimaced, trying not to dwell on the fact that the guy may have strangled a complete stranger.

Thirty minutes later, Caitlyn poked her head through the connecting doorway. "Any word from Detective Ernest?"

"Not yet." He smiled when he noticed she was drinking coffee without her usual flavored creamer. "I'm surprised you're hitting the hard stuff," he joked.

She wrinkled her nose. "Regular cream and sugar make it palatable. I hope we don't have to stay here much longer, or I'll have to find a grocery store to pick up my peppermint creamer."

"I hope so too." He wanted nothing more than for the danger to be over, but he'd miss this. Sharing meals, being with her in the mornings.

Don't go there, he silently warned. *Just don't.*

Thankfully, his phone buzzed. Ernest was up. He quickly called him. "Hey, we have the ID on the victim."

"How?"

Devon explained how Caitlyn had been searching on social media for Snowball's owner when she stumbled across the photograph. "I have it on my phone, here, I'll text it to you."

"Stephanie Phillips," Ernest said once the text had gone through. "Caitlyn is absolutely sure this is our victim?"

"Yes. And we believe Snowball, the cat, may have escaped that night when this guy pulled Stephanie from the car. It explains how chasing the cat aided in Caitlyn's witnessing the murder."

"Yeah, that scenario works," Ernest agreed. "This is

good work, Rainer. I'll call her family, see if they can fill in some of the blanks."

"We could use the name of the killer, for sure."

"Are you still going to have Caitlyn work with the sketch artist?" Ernest asked.

"I am, just in case this guy left with her from Flannery's without knowing her ahead of time."

"Let's hope that's not the way this went down."

"I pray it's not, but you're the one who taught me to keep all options open."

"You might make a good detective yet, Rainer."

Sensing the detective wasn't joking, he smiled. "Thanks. Maybe someday I'll be as good as you."

"Flattery won't get you a recommendation," Ernest shot back. Then his tone changed. "Seriously, we need to stay in touch today. If I get any inkling of a boyfriend who wears size twelve shoes, I'll need Caitlyn to ID him from a six-pack."

"Understood." He caught Caitlyn's blue gaze as she stood petting the cat. "She'll be ready."

"Good. Later." Ernest disconnected from the call.

"You're going to find him today," Caitlyn said from the doorway. "I can feel it."

"Me too." He had to resist the urge to pull her into his arms. "We should consider grabbing breakfast. Once Captain Barstow gets the sketch artist set up, we'll need to go."

"Sounds good, after we give Snowball her antibiotic."

They had the routine down to a science, and soon Snowball had been medicated and fed. Caitlyn gave the feline one last stroke before gathering her purse and heading toward the door.

There was a restaurant within walking distance. Devon

had left his service weapon and shoulder harness in the motel, but he still wore his ankle holster. Since he wasn't in full uniform, he thought the ankle holster would be better in the restaurant. He kept a wary eye out for anyone lurking nearby as they left the room.

Once they were seated in a booth, he noticed Caitlyn switched from coffee to water. After they'd placed their order, she asked, "When do you think the sketch artist will be ready?"

"Hopefully not too much longer, likely around eight or eight thirty." Devon took a sip of his coffee. "Things move faster during the week compared to the skeletal crew we operate with on the weekend."

She nodded. "Same for us at the veterinary clinic."

"Speaking of that, was your boss okay with your calling off work?"

"He wasn't happy until I explained about how I was helping the police find a killer. Then he was very understanding."

Devon hid a wince behind his coffee mug, wishing she hadn't gone into that much detail with the veterinarian. Not that it should matter. He doubted Dr. John, as she called him, would be passing the information along to others. Without Caitlyn or Annette, the clinic would be short-staffed. No reason to worry.

Still, it added a sense of urgency. Once the victim's family was informed of her murder, the news reporters would be flocking to Sevierville like buzzards on a dead rabbit. If it became known that Caitlyn was a witness, his job of protecting her would be five times more difficult.

Unfortunately, there was nothing he could do about that now.

When their food came, he reached over to take Caitlyn's hand. "I'll say the prayer."

"I'd like that." Happiness shone from her blue eyes, and in that moment, he realized she was right about having more life experience. Including faith.

He bowed his head, trying to remember how she'd done it. "Dear Lord, we thank You for this food we are about to eat. We thank You for your guidance as we search for this evil man, and we ask for You to continue to provide us Your strength and wisdom. Amen."

"Amen," Caitlyn echoed. She tightened her hand around his before releasing it. "That was wonderful, Devon."

"Thanks." He cleared his throat and tried not to look around to see if anyone had been watching. There were plenty of believers in the state of Tennessee, but he'd still felt extremely self-conscious. Praying out loud wasn't something he'd ever done.

Sneaking a glance around, he realized no one seemed to notice. Or they didn't care.

"It gets easier with practice," Caitlyn said with a grin.

He flushed. "That obvious, huh?"

"I felt the same way," she confided. "Linc helped me overcome my fears, and God did the rest."

"I'm glad you had Linc guiding you." He knew without being told that Linc would do the same for him if he asked. A comforting thought.

Halfway through breakfast, his phone rang. His heart jumped, but the call wasn't Ernest with good news. It was his partner, Bruce. He debated answering but figured he'd only keep calling. "Hey, Bruce."

"What's with you calling off sick?" Bruce's sharp tone indicated he wasn't happy.

"I'm fine, thanks. How was your weekend?" Devon

asked.

"All I'm saying is that you don't sound sick," Bruce muttered. "My weekend was fine, thanks."

"You and Doreen patch things up?"

"Yeah. But seriously, Dev, what's the issue?"

Devon glanced at Caitlyn. "Nothing much, I just need a couple of days. Personal reasons."

"Huh." Bruce didn't sound convinced. "You know they're going to stick me with some lame rookie."

"You'll get over it." Devon took another bite of his scrambled eggs. "Look, Bruce, I have to go. Be safe out there."

"With a rookie covering my back? Not likely. Later." His partner rang off.

"Who was that?" Caitlyn asked.

"My partner, Bruce. He can be a little . . ." He struggled to find the right word. "Overbearing."

"Do you give him grief when he calls off work?"

"No, but I'm easygoing. Bruce gets tense." He waved a hand. "It's nothing personal, that's just his way."

"I guess you can't pick your partners," she said with a wry smile. "I'm grateful Annette is easy to get along with, both at work and as a roommate."

"Bruce is fine." He brushed off her comments. Cops were brothers in blue, often closer to each other than to their respective families. He could understand why his partner would feel out of sorts. You wanted someone you could trust to cover your back. And a rookie might do his or her best, but it wasn't the same as having a six-year veteran beside you.

They finished eating. He lingered over another cup of coffee, waiting for Captain Barstow to give him the information on the sketch artist.

After he paid the bill, he called the captain but was forced to leave a message. He sighed, then rose. "Ready?"

"Absolutely." Caitlyn slid out of her side of the booth. She was wearing casual clothes, jeans and a pink sweater. He wondered if he'd ever be able to look at her without wanting to kiss her.

Yeah, probably not.

He kept her close to his side as they headed back outside. There weren't many people milling about, which made it easier to make sure there was no one following them.

Back at the motel, he tried his boss again. It was still early, but he'd expected to hear something from the captain by now.

Or from Ernest.

Caitlyn gave Snowball some attention as he paced the length of the motel. How much longer should they wait. Maybe he should simply head out to the police station.

Apparently, patience wasn't his strong suit.

Finally his phone rang. "Hey, Captain."

"Rainer. I have the sketch artist arriving at eight thirty. Can you bring the witness?"

"That works fine, thanks." He tried not to show his relief. "Any word from Detective Ernest?"

"No, other than he's currently meeting with the victim's mother. Hope to hear from him on what he's found out soon."

"Good." Thank goodness the victim had family to inter-view. "I'll wait to hear from him, then."

"I'd like to think I'd hear from him first," Barstow drawled.

"Of course, sir."

"By the way, your partner called in sick today too,"

Barstow added. "You two aren't planning to play hooky together, are you?"

"Not at all." Inwardly, Devon rolled his eyes. Figures Bruce would call off just to avoid being saddled with a rookie. "I'm not going to leave Caitlyn's side, for any reason."

"Just checking."

"See you soon, captain." He disconnected from the call and glanced at Caitlyn. "Did you hear? You're on at eight thirty."

"I'm glad." She offered a wan smile. "I'm anxious to get this over with. Although I'm not sure a picture will help much."

"Your sketch will be key if we find out that our victim didn't know this guy."

"I suppose you're right." She looked disappointed at the thought.

A knock at the door made him jump. Considering he had the Do Not Disturb sign on, he frowned and headed over to the peephole. When he saw his partner standing there dressed in his work uniform, he inwardly groaned and opened the door. "Bruce, what are you doing here?"

"Looking for you." Bruce quickly pulled his service weapon from its holster and leveled it at his chest. "And Caitlyn."

Devon stumbled backward, shocked and dismayed to realize what was happening. The killer was his partner? How was that possible? It didn't make sense. Caitlyn should have recognized his photo on the wall in the police station.

Wait a minute. He thought back, realizing Bruce's photo wasn't there. Because his partner had removed it? Highly likely.

His own partner was going to kill him.

CHAPTER TEN

Caitlyn stared in horror at the man standing in the doorway. She recognized him as the killer. Devon had called him Bruce.

His partner. The murderer was not just a cop, he was Devon's partner.

Keen betrayal was etched on Devon's features. She eased Snowball off her lap and stood. She silently prayed that God would provide the strength and courage she and Devon would need to get away from this madman.

"Don't move. Either of you," Bruce said curtly.

"You're really going to kill us?" Devon asked incredulously. "I'm your partner. We've covered each other's backs for the past three years."

"Hey, I gave you a way out," Bruce said in an exasperated tone. "All you had to do was turn Caitlyn over to me." He glared at her. "This is all your fault. You shouldn't have come through the woods to find us."

Her fault? She couldn't believe his audacity. He was the killer, not her.

"You sent me that email." Devon stared at him. "I have

to say, I didn't suspect you, Bruce. Not for a moment. And I didn't realize you were so tech savvy."

"There's a lot you don't know about me, Rainer." He smirked. "For a guy who wants to be a detective, you're pretty clueless."

She saw Devon's fingers tighten into fists, but he didn't move. Or show any reaction on his face.

"Why did you kill Stephanie?" Caitlyn was surprised at how calm her voice sounded. She wasn't afraid to die, knowing she'd be in a better place with God, but at the same time, she didn't want Bruce to get away with murder.

And right now, she and Devon were the only ones who knew the truth. If they disappeared and turned up dead, she didn't believe Detective Ernest would ever pinpoint Bruce, a cop from their own precinct, as the murderer.

"You figured out her name, huh?" Bruce narrowed his gaze. "I guess it's a good thing I got here when I did."

"And how did you find us?" Devon asked.

"Again, you underestimate me, Rainer. Tracing your cell phone was easy, I was able to get here in time to see the two of you walk over from the restaurant. Piece of cake."

Devon's expression tightened at his words.

"Well, if you're going to kill us, you may as well fill in the blanks about why you killed Stephanie," Caitlyn said. "I know you're married, so I'm assuming Stephanie was your mistress."

"Hear that, Rainer?" Bruce's tone goaded his partner. "She'd make a better detective than you."

"What happened, did your wife find out about your cheating?" Devon asked, ignoring the jab.

"No, Doreen has no idea about my liaison, which is how it should be. Unfortunately, Steph was stupid enough to get pregnant." He shook his head in disgust. "She expected me to pay child support, can you believe it? Her getting pregnant wasn't supposed to happen."

Caitlyn felt even worse, knowing that Stephanie was pregnant when he'd killed her. Three people. Stephanie, her unborn baby, and Nate Powers. Bruce had killed three people without hesitation.

She had no doubt he'd murder her and Devon without blinking an eye.

"You killed your mistress because she got pregnant?" Devon asked incredulously. "And why is that her fault? It takes two people to create a baby. Besides, any decent man would take responsibility for his actions."

Anger tightened Bruce's features. Caitlyn hoped Devon hadn't pushed him too far. She subtly searched the room for anything to use as a weapon. She wasn't far from the ceramic lamp on the table, but the heavy object wouldn't be much use against a gun. Still, she subtly edged closer, feeling the need to do something.

"Shut up," Bruce snapped. "She should have taken care of it. Not come crying to me about needing money to help raise the brat. And if Caitlyn hadn't seen us, no one would have ever figured out what I'd done."

"Don't be too sure of that," Devon said. "Detective Ernest is speaking to Stephanie's family right now. They'll sell you out in a heartbeat."

"They know nothing," Bruce scoffed. "Steph knew better than to tell her family she was sleeping with a married man." Then his features hardened. "And I told you to shut up."

As the two men were talking, she'd managed to edge closer to the lamp. But she was afraid to make a grab for it. Devon was too close; if Bruce shot him, he'd die.

Her heart squeezed painfully in her chest as she realized how much she loved him. She couldn't lose him today. Not now. And certainly not to this coldhearted killer.

Please, Lord, give me strength!

"You keep telling me to shut up, but what exactly is your plan?" Devon asked. "Are you going to shoot us here at the motel? It's daytime, someone will hear the gunfire."

"Don't worry, my plan is to take the two of you someplace where no one will hear the gunfire." Bruce grinned in a way that reminded her of the Preacher. "Have you noticed I'm in my uniform? No one will question me taking the two of you out of here in cuffs." As he spoke, he reached toward the handcuffs on his tool belt.

Now!

It seemed as if Devon may have heard the tiny voice in her mind because he moved at the exact same time she did. Caitlyn grabbed the lamp and threw it at Bruce in one smooth motion. Devon dove to the side doing some sort of martial arts move with his foot to kick at Bruce's gun. There was a screeching *meow* as Snowball jumped from the top of the dresser, landing on Bruce, her claws raking down his face.

Bruce screamed in pain and fury. He tried to deflect the lamp with his free hand, even as he dropped his gun from the force of Devon's kick. Deep claw marks bled profusely down his face. Caitlyn grabbed the chair and tossed it at Bruce as well.

"No!" Bruce shouted, but it was too late, Devon had pulled his weapon from his ankle holster and leveled it at his partner.

"Put your hands up or I'll shoot!" Devon said.

It made Caitlyn nervous because she and Devon were dressed casually, while Bruce was in full uniform. Who would believe that Devon was the good guy and not Bruce?

Bruce began to put his hands up, but then he reached for his utility belt. Seeing the Taser there, she froze, staring in horror.

Then Devon fired one shot, hitting Bruce in the center of his chest. The guy looked stunned as he fell back against the door. Bruce put a hand to his heart as if the organ may have stopped cold. Caitlyn was shocked Devon had shot him, but then she realized Bruce was still conscious and that there wasn't any blood.

What in the world?

"I know you're wearing body armor," Devon said calmly. "Make another wrong move and I'll go for a head shot."

Bruce grimaced, his hand still pressed to his chest. "Hurts," he gasped.

"Not as badly as you hurt Stephanie," she shot back.

"Grab his handcuffs, Caitlyn, but stay to the side so that I have a clear shot at him," Devon directed.

She did as he asked, handing the cuffs to Devon.

Bruce closed his eyes without moving as if realizing defeat. Between her, Devon, and Snowball, they'd gotten the better of him.

Devon expertly snapped one silver bracelet around his wrist. Then he pushed Bruce face down on the floor and wrenched his other arm behind his back to cuff his two hands together. The cop groaned loudly.

"Bruce Whitmore, you have the right to remain silent. Everything you say can and will be used against you in a court of law . . ." Devon continued reading his former partner his rights as he forced the guy to his feet.

Caitlyn took a step back, her hands shaking in the aftermath of the scuffle. Noises from outside the motel indicated people were running for cover because of the gunfire. She had no doubt the police would arrive soon.

Snowball meowed again and leaped into her lap. She cuddled the cat close, wishing she could stop trembling.

It was finally over.

DEVON PUSHED Bruce into a chair and called Detective Ernest who answered immediately. "No leads from Stephanie Phillip's family," Ernest said in lieu of a greeting.

"I have Bruce Whitmore in custody. He killed Stephanie because she was pregnant with his child. He tried to take me and Caitlyn away at gunpoint to kill us too."

"Whitmore?" Ernest echoed incredulously. "Your partner? Are you sure?"

"He confessed," Devon said firmly. "Caitlyn is my witness. Snowball is Stephanie's cat, and she clawed him good. Besides, you know we suspected the perp was a cop. We were right about that."

Ernest whistled. "I'll be there in ten."

The wail of sirens could already be heard. "I had to shoot him. Thankfully, he was wearing body armor, but someone called it in. The cops are on their way."

"Hang tight," Ernest said. "Wait, did he tell you where Stephanie's body is?"

Devon glanced at Bruce who had wisely remained silent once he'd been placed in cuffs. "No."

Ernest muttered a curse. "Okay, I'll be there as soon as I can."

"Thanks." Devon slid his phone into his pocket and glanced at Caitlyn. She sat on the edge of the bed, looking shell-shocked. He was proud of her for throwing the lamp at Bruce and more than a little surprised by the fact that they both took action at the exact same time. Yet he also knew these were hardly normal circumstances. "Caitlyn? Are you okay?"

"Yes." Her voice was faint. Her blue eyes clung to his. "I don't understand why his photograph wasn't up on the wall of the police station."

"He must have removed it." Devon glanced at Bruce. "He was afraid you'd recognize him."

"I would have." She shivered and shook her head. The cat purred in her arms as if in agreement. "It's just hard to believe the danger is finally over."

"You're safe now." Devon was thankful they'd escaped unscathed. Especially Caitlyn. He shifted his gaze to his former partner. He and Bruce had been partners for three years. They hadn't been super close, though, because he'd never appreciated how Bruce had talked down about his wife, Doreen. Thinking back, he wasn't entirely surprised to know Bruce had cheated on his wife.

But to think the guy had been capable of this? Of strangling a woman with his bare hands? Killing an innocent man, then hunting him and Caitlyn down?

He had to wonder about his instincts, or lack thereof. To know he'd been so far off base . . .

"Don't, Devon." Caitlyn came over to stand beside him. "I can tell you're wondering how you missed this, but it's not your fault. It's Bruce's fault. He betrayed his marriage vows and the promise to protect and serve. How could you know he'd do something like this?"

"There must have been signs . . ."

"No, Devon. The devil is good at deception and trickery." She reached for his hand. "Think about how blessed we were to have had God's strength and support. We triumphed today, not him."

A hint of a smile tugged at the corner of his mouth. "You're right. We did have God on our side."

Bruce snorted. "Give me a break," he muttered.

Devon ignored him, gazing into Caitlyn's blue eyes. "I'm thankful you weren't hurt."

"Ditto," she said with a smile. Snowball meowed.

He wanted nothing more than to pull her into his arms, but the sirens were right outside the motel, so he reluctantly pulled away to open the door. He held his arms up, his gun pointing toward the ceiling. "I'm Officer Devon Rainer, and I have another officer, Bruce Whitmore, in custody!"

"Rainer?" His fellow officer Tim Cast gaped in surprise. "You arrested your partner?"

"I did. Bruce confessed to killing Stephanie Phillips. Caitlyn Weston witnessed the crime, so he tried to kill both of us too." Devon didn't lower his hands, not wanting to give any of the officers standing there a reason to panic and fire. "Detective Ernest knows the entire story and is on his way."

Another cop, Officer Hill, approached from the side. "I need your weapon."

Devon didn't hesitate to hand it over. The ankle gun was a smaller caliber weapon, which was one of the reasons he'd taken the shot at Bruce even at close range. "Whitmore is wearing body armor, you'll find the mangled slug from my weapon embedded in it."

"Understood." Once Hill had taken his gun, Devon lowered his arms. "I'll need you and the witness to stay separated while we gather information."

"I know." Devon knew how it was done. "But keep in mind Caitlyn is watching over Stephanie's cat, Snowball.

You'll notice the claw marks on Bruce's face, I'm thinking the feline wasn't very fond of him."

"I've heard animals can be perceptive," Cast said, joining them. "Hill, you take Dev. I'll take Caitlyn."

"Got it," Hill agreed.

The interviews took longer than he'd expected. Although he knew that discharging his firearm added a higher scrutiny to the process.

Ernest arrived and took over the investigation. Devon had to start over again, but he was glad Ernest believed him. Hill and Cast took Bruce to jail while he and Caitlyn were being questioned.

Hours later, he and Caitlyn were finally free to go.

He placed a hand under Caitlyn's elbow. "I'll drive you home."

"What about the cat supplies?" She dug in her heels. "I don't want to buy all new stuff."

The shooting had taken place in his room, and while the connecting door was open, there wouldn't be any kind of evidence in there. He crossed over to Ernest and asked for the items.

"It's a crime scene," Ernest said.

"Technically, it's not. Besides, Snowball helped save the day. The way she launched herself at Bruce helped me disarm him."

Ernest sighed. "Fine. You can take the cat supplies, but nothing else."

"That's all we need, thanks." Devon moved toward the motel room.

"Rainer?" Ernest called.

He turned. "Yeah?"

"Nice work today." The detective gave him a nod, then turned away.

Devon gathered all the cat supplies, wondering if that was Ernest's way of giving him a recommendation for a promotion to detective. Then he thought about all the signs he'd missed related to Bruce being their perp and told himself he probably shouldn't bother taking the exam.

No matter what Caitlyn said, Devon knew he should have sensed something was off about Bruce. How could he have not realized the man was a cold-blooded killer?

The devil is good at deception and trickery.

Caitlyn's words echoed in his mind. He wondered if that was something she'd heard from attending church.

Was she right? Maybe it wasn't all his fault for not seeing the truth about his partner. He gave himself a mental shake.

One thing was for sure, he had a lot to learn about God, faith, and the Bible.

When he had everything stored in the unmarked SUV, he drove Caitlyn back to her apartment. He pulled in next to her Honda where she'd left it, was it only thirty-six hours ago?

She sat staring through the windshield without moving, the cat still curled in her lap.

"Is something wrong?"

"I don't want to be alone." The words were soft. "Please, Devon, will you stay here with me? Or we can go to your house. I just—don't want to be alone."

"I'll stay with you," he assured her. "And I won't leave until you're ready."

She smiled wanly and pushed out of the car. He carried in Snowball's things, waiting for her to unlock the door.

Inside Caitlyn's apartment, he stared at her tree for a long moment. He never bothered to decorate, but now as he looked at the star she'd perched on the top, he truly understood its significance.

The star of Bethlehem had shone the way to Jesus.

Just like Caitlyn had shown him the way to God.

He stood for long moments as she took care of the cat. Then she crossed over to join him. "I wanted to thank you again, Devon, for everything you did for me."

He turned to face her. "You held your own. I was shocked when you threw the lamp at him."

"I was surprised that Snowball attacked him." She slid her arm around his waist and leaned against him.

"The cat did her part," he agreed. He pulled Caitlyn closer into his arms. "We made a great team."

"With God's help."

"Yes, with God's help." He inhaled the lavender scent that seemed to linger in her hair. "Caitlyn, I'd like to see you again, once this is over. Maybe we could, uh, get together for dinner sometime?"

It took a moment for the words to sink in. Then she pulled away to face him. "Dinner sounds great, Devon. But do you mind if I ask what's changed?"

He frowned. "What do you mean?"

"Two months ago, I asked you to get together for coffee and you declined."

He winced. "I remember."

"So what has changed?"

He tried to think of a way to explain his feelings. "I've fallen in love with you."

Her jaw dropped, her eyes widening incredulously. "You what?"

"I love you." He smiled and lightly stroked her cheek. "I was an idiot for turning you down, Caitlyn. You're smart, beautiful, brave, and kindhearted. I completely understand you'll need more time, but I just wanted you to know the truth about how much I love you."

"Oh, Devon, I'm glad to hear you say that." She laughed and stepped into his embrace. "Because I love you too."

"Are you sure?" He stared down into her eyes. "You're several years younger than me. And the circumstances of our being together these past two days are hardly normal. We can take things slow, as long as you let me be a part of your life."

"Do you think Linc and Jayme took things slow?" she teased.

"No. But they're older . . ." He stopped when she narrowed her gaze.

"Don't treat me like a kid, Devon. I lived with an abusive Preacher, escaped, lived on the streets while keeping away from many a predator. I'm not young, Devon. Not in any way that counts."

"I know you've had a difficult upbringing," he acknowledged. "That's the reason I wanted to take things slow. To make sure you understand what you're getting into."

"I know my heart," she said simply. "And I love you, Devon. These past two days have only given me more reasons to love you."

"I'm not sure what I did to deserve you," he murmured.

"Maybe it's more that I deserve you," she countered. "God brought us together for a reason, Devon. Who are we to question His will?"

"Not me," he agreed. He pulled her close and kissed her, pouring his heart and soul into the embrace.

Maybe God believed they deserved each other.

EPILOGUE

Christmas Eve

Caitlyn ran a hand down her red sweater dress before letting Devon into the apartment. They'd agreed to meet here prior to heading over to Linc and Jayme's for dinner. She had agonized over what to get him as a gift, and in the end, she'd simply framed a picture she'd taken of the two of them at the winter carnival in Knoxville.

It wasn't enough, but she'd wanted something to show him how perfect they were for each other. Despite the way they'd pretty much seen each other every day for the past two weeks, she knew he still wanted to take things slow.

On one hand, she was grateful he wasn't pushy, but on the other hand, it irked her that he made a big deal of their five-year age difference. As if that mattered.

"Hi," she greeted him.

"Hi. You look fantastic." He came inside and gave her a quick kiss. She noticed he wasn't carrying a gift, which was fine with her as her picture wasn't anything expensive. And honestly, it took the pressure off.

"Thanks." She closed the door behind him.

"I have news." Devon took a seat on the sofa and patted the cushion, indicating she should sit beside him. "They found Stephanie's body."

She sucked in a quick breath and sat beside him. Snowball jumped into her lap, and she stroked the kitten. "Wow, I'm happy for her family. At least they'll have some closure."

"Yeah, and believe it or not, Bruce's fingerprint was lifted from her neck." He shook his head. "Looks like he might confess in an attempt to get a lighter sentence."

"Lighter sentence?" She frowned. "I don't like the sound of that."

"Oh, he'll still do life in prison, but they'll likely take the death penalty off the table."

She sighed. "I guess I'm fine with that, then."

"Even better, you won't have to testify against him." Devon wrapped his arm around her shoulder and pulled her close. "I know you'd do great on the stand, but it's an added stress you don't need."

"Yeah." She rested her head on his shoulder. "I'm glad it's over."

"Me too." He kissed her temple, then shifted on the sofa. "Oh, and Stephanie's family is really glad you're willing to keep Angel, a.k.a. Snowball, as Stephanie's mom is allergic."

"Snowball is an angel in that she led me to Stephanie that night." She looked down at the cat. "If not for her, we may not be sitting here right now."

"Oh, we would have eventually," Devon said. "I think God would have found another way for us to get together."

Devon had joined her at church last weekend, and she'd been thrilled at how closely he'd paid attention to the pastor's words. Linc and Jayme had come too, and she noticed Linc had been impressed as well.

"You're right," she agreed. "I, uh, have a small gift for you."

"You do?" He sounded surprised. "I have one for you too."

"Really?" She sat up and turned to look at him. Then she saw the small velvet box in his hand. Her heart stuttered, and she blinked to make sure she wasn't imagining things.

"Caitlyn." Devon slid off the sofa and knelt before her. He opened the box revealing a modest, sparkly diamond ring. "Will you do me the honor of marrying me?"

"I—yes!" She was stunned by his proposal. She wrapped her arms around him and kissed him. After a long moment, they came up for air. He slid the ring onto the fourth finger of her left hand. She couldn't seem to think of anything to say. "You took me by surprise," she finally admitted. "I never expected you to propose."

"I talked to Linc yesterday, to get his blessing." Devon smiled. "I know he's not your father, but he's your older brother, and his opinion matters."

"Oh, Devon." She was touched that he'd asked Linc for his blessing. "That's the sweetest thing anyone has done for me."

"The first of many," Devon promised. "I plan to spend the rest of my life making you happy."

"And I will do the same." She kissed him again, then rose to get her gift. Feeling a bit self-conscious, she gave it to him. "Merry Christmas, Devon."

He unwrapped the gift, gazing in wonder at the framed picture of them. "This is perfect, Caitlyn. Thank you."

"You're welcome." She glanced at her diamond ring. "It doesn't seem like enough after your proposal."

"It's more than I expected." He pulled her close and kissed her again. "I love you. Merry Christmas, Caitlyn."

"I love you too." She sighed and silently thanked God for bringing Devon into her life.

DEAR READER

I hope you've enjoyed my Smoky Mountain Secrets series. I had this idea many years ago, and I'm thrilled that the entire series is complete. Thanks again to all of you who took the time to let me know how much you've enjoyed these books.

If you enjoyed this story, please consider taking a moment to leave a review either from the site where you purchased the book or on Goodreads. Reviews are very important to authors, and I would really appreciate your kind gesture.

I'm truly blessed to have such wonderful readers! Please know I'd love to hear from you! I can be found on Facebook at https://www.facebook.com/LauraScottBooks, on Twitter at https://twitter.com/laurascottbooks, and on Instagram at https://www.instagram.com/laurascottbooks/. I can also be reached through my website at https://www.laurascottbooks.com. If you're interested in hearing about my new releases, consider signing up for my newsletter. You'll receive a free novella that is not available for purchase through any platform. It is exclusive only to my newsletter subscribers.

Lastly, if you haven't read my Security Specialists, Inc. series, you may want to give it a try. These are action packed stories. I've included the first chapter of *Target For Terror* for you to sample. I'm currently working on the fourth book in the series, *Target For Treason*.

Until next time,
Laura Scott

TARGET FOR TERROR

June 30 – 7:06 p.m. – Washington, DC

"I saw the man who shot me."

Natalia Sokolova heard the words in rapid Russian, the language of her birth, and glanced in surprise at her patient. Her fingers faltered in the process of hanging the second unit of blood on his IV pump.

"Are you sure?" she responded in Russian.

Josef Korolev nodded, his gaze boring into hers. "The FBI agent in the front row of the crowd shot me."

What? An FBI agent shot him? "That can't be. You must be mistaken."

"*Eto ne oshibk.* There was no mistake." The deputy prime minister of Russia's voice was hoarse, scratchy because of the breathing tube that had been removed just a short while ago. "I saw him."

She finished connecting the unit of blood, then placed a reassuring hand on his arm. The deputy prime minister had come to Washington, DC, to deliver a speech regarding the importance of the Middle East Peace Summit to take place at the International Conference scheduled in Moscow.

During the speech he'd been shot, which had caused quite the international incident. The entire country was in an uproar over the event.

Josef Korolev grabbed her hand as if willing her to believe him. "Listen to me. I saw him. The bullet came from close range. They didn't think I'd live long enough to tell."

He spoke with such heartfelt conviction the ugly shadow of doubt was difficult to ignore. Especially given the tenuous relationship between the US and Russia. Could her patient be confused? Thinking back over the few hours since Josef had returned from the OR, she counted the amount of narcotics she'd given him. A total of ten milligrams of morphine wasn't too much for such a large man over a three-hour period. Obviously, Josef Korolev had been shot, the Secret Service agents standing guard outside his room substantiated that. But by an FBI agent? She seriously doubted it.

The rumor zipping through the hospital grapevine claimed some sort of disgruntled assassin from the Russian mob was the prime suspect in the shooting because the Russian government was losing its battle against the Mafia underworld. The hospital had cops stationed at each entrance and several more in the ICU.

Yet all the police in the world couldn't keep Korolev safe if the suspect was really an FBI agent.

No, she couldn't believe it. Likely, the morphine was too much for him.

"You're safe here," she assured him. "This is the surgical intensive care unit at Washington University Hospital, and I'm your nurse, Natalia. There are two men from your country here with you, but they've stepped out to discuss arrangements for your transfer home. They'll be back soon."

"I am not safe. Will never be safe, not until I get back to

Russia. You must help me." He was old enough to be her father, but his grip was strong as he clung desperately to her hand. "Do not leave me alone with anyone from the American government."

"I won't." She leaned over to pull his blanket up over his chest, knowing the blood transfusion would give him a chill.

"Where did you get this?" His gaze zeroed in on the two pendants dangling from around her neck. His heart rate jumped up, causing his monitor to alarm overhead, but she ignored the noise when he touched the crescent-shaped moon with the three amethyst stars pendant that hung above her Christian cross. "Who gave this to you?"

"My mother." The treasured pendant was the only item she possessed from her Russian birth mother. She would have explained more, but Josef had grown agitated, muttering something she couldn't quite make out, so she tucked the pendants underneath the collar of her scrubs. "Shh. Relax now."

Before Josef could say anything more, she noticed a trio of official-looking American men walking toward her patient's room. Natalia straightened, watching through the glass window as they paused to speak with the Secret Service agent and the Russian countrymen before entering. Instinctively, she took a step closer as if her mere presence would save Josef from harm.

"The deputy prime minister is awake?" The shortest of the three, a man whose FBI name tag identified him as T. Saunders, pinned her with a sharp gaze. "The breathing tube is out? He's able to answer questions?"

"Yes." She winced as Josef's hand grabbed hers and squeezed tightly to the point of bringing pain. She tried to smile. "Well, not really," she amended. "He's confused, speaking gibberish. The medication has strongly affected him."

The iron hold on her hand didn't ease at her words because Josef did not understand English. She bent close to her patient, speaking in Russian, "Shh, relax. I explained how you are too confused from the narcotics to discuss anything about the shooting right now."

Josef's grip on her hand subsided, and he closed his eyes, feigning sleep.

"You speak Russian too?" Saunders was clearly the leader of the three. The sneer on his face suggested he accused her of something vile.

"Yes, but only a little." Natalia forced a smile. "My adopted mother had dual citizenship in both Russia and here in the US. She taught me some basics of the language. But I'm afraid I have grown very rusty over the years since she passed away." After her adopted mother died, she continued to practice her Russian with her friend Ivan. Yet she couldn't explain the deep-seated certainty that she needed to downplay the extent of her knowledge of the Russian language.

"What exactly did he say to you?" Saunders persisted.

"I don't know, the words didn't make sense. I told you, he's confused. I've given him about ten milligrams of morphine."

"Maybe we need one of our own interpreters in here to tell us just how confused he is." Saunders's gaze challenged her opinion.

She schooled her features not to show her annoyance. "Please do. There is a Russian interpreter, Ivan Rasacovich, available through the hospital social service department if you are interested. As I mentioned, I am not an expert."

"It's a miracle he's survived the shooting," the tallest of

the three commented as if to change the direction of the conversation and ease the tension in the room. He was the only one without an FBI name tag, his badge simply read visitor.

"Yeah, no thanks to you, Dreyer," Saunders said in a snide tone.

Dreyer's mouth tightened. "Or to any of us assigned to protect him. The whole event was a debacle, and we can't even get a clear view of the shootings from the cameras." His gaze swept back to the patient. "He's lucky the bullet missed his heart and only grazed his lung."

Natalia doubted Josef Korolev felt very lucky considering the lengthy surgical incision extending around the side of his chest, but she held her tongue. As the FBI agents argued about who might have wanted to silence the deputy prime minister, the man named Dreyer stared at her to the point he made her uncomfortable. He was younger than the others and not dressed in the traditional FBI garb consisting of black suit and matching tie. The simple black T-shirt, stretched across his taut chest, dark jacket, and slacks looked far too casual for an official visit. He might be described as handsome if you liked tall, dark-haired American men with square jaws and brilliant blue eyes.

Good thing she didn't. A broken heart had cured her of such foolishness. She preferred peace and quiet in her life compared to the never-ending drama of a so-called relationship.

"Please, I must ask you all to leave. It's time to change his dressing." Another blatant lie, but she didn't care. She didn't know who these men were or which person she could trust. There was no reason on earth for the FBI to shoot a Russian diplomat, especially while he happened to be ensconced in their protection, but she couldn't totally discount her patient's accusation either. Something wasn't quite right with this picture, and she wanted these men to go away because she didn't have time to waste figuring it out.

Her job was to provide nursing care to her patients. When the men simply stood there, she scowled. "The chest dressing needs to be changed. Would you rather I risk infection?"

"No, of course not." Dreyer nodded as if he completely understood. "We'll give you the privacy you need. Gentlemen?" He gestured toward the door.

The other two sent her hostile looks but, eventually, turned and left the room. Dreyer lingered. She tried not to squirm under his intense gaze.

"Natalia, do you also understand Ukrainian?" His attempt at the language was stilted at best, and his tone was low as if he didn't want anyone to overhear.

At first she was suspicious when he used her name, then she realized she was wearing her hospital ID badge. She sniffed. Someone should tell the man his accent was terrible. "Not fluently, but enough to understand the basics," she answered in Ukrainian. Russian was very different from Ukrainian, a fact he should know. "Why do you ask?"

"No reason in particular. Good night, Natalia." The man turned and followed the others out of the room, leaving her to wonder what test she'd just taken.

And if she'd passed—or failed?

JUNE 30 – 9:18 p.m. – Washington, DC

. . .

SHE DIDN'T GET much time alone with Josef from that point on, mostly because she was running back and forth helping one of her co-workers whose patient took a sudden turn for the worse. She had wanted to ask Josef about her pendant though, since he acted as if it were familiar. Had he seen one just like it? Did he perhaps know who designed the dainty Russian jewelry? The idea her pendant might hold a clue to the true identity of her birth parents filled her with excitement. She couldn't wait to ask him more.

Natalia sat down at the workstation outside her patient's room to make a quick notation in Josef's chart when a shrill triple-beeping sound grabbed her attention. Her gaze snapped from the computer to the blinking red alarm on the central bank of monitors.

Josef's heart rate had gone into a life-threatening arrhythmia.

"Call a code blue." Natalia jumped from her seat and dashed into the room, knocking aside the wide-eyed guard in her haste to reach her patient. V-tach without a pulse. She crawled up to kneel on the side of his bed to perform chest compressions.

The room flooded with medical staff who shoved the hovering Russian countrymen out of the way. Someone took over CPR while another began to give breaths with an Ambu bag.

"Shock him," the surgeon snapped.

Natalia was already slapping the defib patches onto his chest. She connected them to the defibrillator and twisted the knob to 200 joules. "All clear?" she asked before she administered a shock.

"Shock him again," Dr. Ventura ordered. She did as she was told, knowing the algorithm.

"Give him a third shock." All eyes in the room were on

the monitor, hoping and praying for a conversion into a normal heart rhythm. But that didn't happen. "Continue CPR and give a bolus of amiodarone," the surgeon ordered. "Then start a drip."

Natalia injected the medication. She grabbed the drip prepared by the pharmacist on duty and hung it on the IV tubing. The surgeon continued to shout orders for labs, for more medication, and to place a breathing tube. They shocked the patient again, then one last time after another round of CPR. As a team, they labored over the patient for a good forty minutes, but despite their best efforts, his heart went into asystole.

Dr. Ventura, the surgeon who'd operated on Josef Korolev, finally shook his head and raised a hand. "Stop CPR. There's nothing more to do."

No! How had this happened? Her pulse raced from the adrenaline coursing through her system, yet she could only stare at the still, peaceful features of Josef Korolev. Less than three hours ago he'd spoken to her.

Now he was dead.

She put a hand to her chest as if to ease the pressure there. Dear Lord, she couldn't believe he was *dead.* Loud vocal protests from the Russian countrymen in the room ricocheted off the walls. Words of comfort failed her.

What could she say to them? She didn't know what had happened. Why had a stable patient suddenly gone into V-tach? Was there something she'd missed? She had several years of nursing experience, but no one was perfect. Yet, in retracing her actions during her shift, she could not think of a single thing she would have done differently. The only oddity, other than all the political hoopla of taking care of a VIP, had been the strange visit from the three Americans.

And Josef's wild accusations.

FBI and Secret Service agents swarmed the room. She stayed close, listening as they grilled the poor guard who'd been left on duty. When they'd finished with the guard, they turned their attention to her.

"Who else has access to this room, besides you?" the FBI agent named Wilcox asked in a harsh tone. "Was he left alone at all?"

"Most of the time he wasn't alone, the Russian countrymen were with him." Natalia glanced nervously between the two men. Why were they asking these questions? Did they honestly suspect something had been done to Josef intentionally?

Should she tell them what Josef had claimed? Yet, if his off-the-wall accusations were true, would telling the FBI help? Saunders and Bentley, the two men who'd come in with Dreyer, hadn't seemed at all sympathetic. She could easily imagine them protecting their own, especially given the political ramifications.

"Yeah, but they spent most of their time in the corner talking in low tones," the guard who'd been on duty said with disgust. "As if I could understand what they babbled about."

"Any member of the hospital staff would have access to his room," Natalia added. "But none of them would have come in unless it was to troubleshoot an alarm or respond to a call light. This isn't the first VIP we've taken care of."

"She wasn't in the room for the past forty minutes," the guard bailed her out. "The only person to come into the room was the housekeeping guy who cleaned the room." When twin sets of cold eyes stared at him, the guard shrugged. "Hey, I took his name—Ray Johnson—it's on the list. Johnson left the room about five minutes before the monitor alarmed."

The FBI agent turned back to her. "Do you know this housekeeper named Ray Johnson?"

"Of course I do." She frantically searched her memory, but she couldn't honestly remember seeing him at any time during her shift. A chill slid down her spine. "I don't remember seeing him around, but I was busy."

The two men exchanged a long glance. "What does he look like?"

"Tall, close to six feet, and very thin. Black hair, usually uncombed." She shrugged. It wasn't as if she knew the man on a personal basis, just enough to say hello.

"Caucasian?"

She nodded.

The guard frowned. "Sounds right, although I didn't think he was that tall, closer to five-ten or eleven. And his hair was more brown than black."

Wilcox, the FBI agent, turned toward the hospital security guard. "Pull up Ray Johnson's picture."

"I tried, but there's something wrong with the computer program." The security guard hunched his shoulders at their incredulous look. "They're working on it."

Wilcox muttered a curse.

Tiny hairs along her nape lifted in warning. *Stop it.* She gave her head a hard shake. Too many crime novels had gone straight to her head. There was no reason to let Josef's outrageous suspicions make her see something sinister in what had caused the code blue. Josef was in his late fifties, and he obviously had a bad heart. She'd noticed minor changes in his EKG after surgery, indicating a possible mild myocardial infarction. The surgeon had ordered serial cardiac enzyme tests to see how much damage his heart had sustained.

Obviously more than they'd realized.

The ruckus didn't settle down for a long while. After she'd given her story to what seemed like dozens of agents, they allowed her to leave. As she gathered her scattered paperwork into some semblance of order, she listened to the ongoing debate.

Some of the problems were simple, like who should talk to the media at the press conference. But others were more difficult, such as who would oversee Josef Korolev's autopsy —someone from the US government or Russian medical examiners? After a heated argument, it was decided both sides would be present during the procedure.

But by the time she was able to leave, a couple of hours after the code blue, she had not seen any of the three men who'd stopped in to visit Josef's room earlier that evening. She'd mentioned it to the agents, and their names were on the guard's infamous list.

Their glaring absence, in the face of Josef's death, gnawed at her on the way home.

Natalia fingered her moon pendant and cross as she rode the red Metro train, grateful for its calming effect. Had she imagined Josef Korolev's odd reaction to her pendant? Had he recognized it, or was it similar to something he'd seen before? Maybe the design, the symbol of a crescent moon with three stars on one end, was specifically Russian in nature. She could easily picture a little shop in Moscow where dozens of pendants just like hers were sold.

She'd concentrated on finding her birth mother through deciphering her adopted mother's journal. She'd never considered the necklace design in itself to be important.

Josef's reaction convinced her that the moon-shaped pendant was a clue. Her chest swelled with anticipation at the thought. She knew her birth mother was Russian, as her adopted mother was. Other than the first name of Anya, she didn't know anything else about her birth mother. Since her adopted mother had passed away three years ago, she'd intensified her search for her roots. Had become obsessed with knowing who she was and who the woman was who'd given her away. Natalia had collected as much information as she could about Kazan, the city of her birth.

As soon as she got home, she'd add this new puzzle piece to the other bits of information she'd gathered about her birth mother. Maybe researching the pendant on the internet would reveal something she'd missed.

A young man shuffled past her, pressing purposefully against her. She dodged him, slipped a hand into her canvas bag, and closed her fingers around the can of pepper spray hidden in there. The crime rate was notorious in DC, and she was jittery enough after the strange events during her shift.

The guy took a seat across from her on the other side of the train. She took a deep breath, then let it out slowly, but she didn't relax her grip on the pepper spray. Illegal or not, she intended to use the weapon if needed.

Ignoring the man staring at her, she trained her gaze on a spider web in the corner of the car. A fly caught in the silken strands fought to get free.

A shiver rippled along her nerves, and she glanced away. Anxious, she waited for her stop to come up. She wanted nothing more than to get home.

Stepping off the train at her Brookland stop, she hurried up the stairs to the main level, staring straight ahead but screening the people around her from the corner of her eye. The events at the hospital had put her on edge. For a moment, she thought the man coming up on her left side looked familiar, but then he turned and headed off in the opposite direction.

Ridiculous to be so paranoid. She gave herself a mental shake. She was a critical care nurse. It wasn't as if she were the one in danger.

Not like Josef Korolev had been. Someone hated him or what he represented enough to shoot him. To ultimately cause his death. Although the Cold War had been declared over decades ago, the animosity between Russia and the US hadn't changed much. If a member of the Russian Mafia was the killer, why wait until Josef was on US soil to finish him off?

Nothing made sense.

She used her rideshare app to get a ride. It didn't take long for a guy in a black sedan to drive up next to her. "Are you Natalia?"

"Yes, thanks." Her house was only a dozen blocks from the Metro station, but she didn't care. She'd pay him extra for his time. "My address is eleven forty-one Girard Street."

The driver grunted, shrugged, and pulled into traffic. Natalia dropped her head against the back of the seat. Thankfully, she was almost home. She considered having a dish of ice cream to relax after her wretched day. At least tomorrow was Friday and she had the weekend off.

"Here ya go, lady."

She opened her eyes, realizing the rideshare had pulled over to the curb. He'd stopped in front of the house next to hers, but she wasn't in the mood to argue.

"Thank you." She added a tip to the rideshare fare via the app on her phone, then opened the door, swung her legs out, gathered her canvas bag, and stood.

Ka-boom! The blast rocked the stillness of the night. The ground trembled beneath her feet. She stumbled and fell backward into the sedan.

The windows burst outward in a hailstorm of glass.

Shards rained down against the car like bullets from a machine gun as a second explosion shattered the night.

It took her a moment to realize it was her house, the precious home she had once shared with her mother, engulfed in roaring flames.